D1229691

From the

LUCY LARCOM BOOK FUND

Made possible by generous gifts to the
WHEATON COLLEGE LIBRARY
Norton, Massachusetts

Virtual Rivers

Virtual

Rivers

Lessons from the Mountain Rivers of the
Colorado Front Range

ELLEN E. WOHL

YALE UNIVERSITY PRESS   NEW HAVEN & LONDON

Copyright © 2001 by Yale University.
All rights reserved.
This book may not be reproduced, in whole or in part, including illustrations, in any form (beyond that copying permitted by Sections 107 and 108 of the U.S. Copyright Law and except by reviewers for the public press), without written permission from the publishers.

All photographs not otherwise credited were taken by the author.

Designed by James J. Johnson and set in E+F Scala type by Tseng Information Systems, Inc. Printed in the United States of America.

Library of Congress Cataloging-in-Publication Data

Wohl, Ellen E., 1962–
Virtual rivers : lessons from the mountain rivers of the Colorado front range / Ellen E. Wohl.
    p. cm.
Includes bibliographical references (p.).
ISBN 0-300-08484-6 (cloth : alk. paper)
1. South Platte River Watershed (Colo. and Neb.)—History.   2. Water use—South Platte River Watershed (Colo. and Neb.)—History.   3. Rivers—Colorado—History.   I. Title.
GB1227.S67 W65   2001
333.91'6214'097887—dc21   00-043771

A catalogue record for this book is available from the British Library.

The paper in this book meets the guidelines for permanence and durability of the Committee on Production Guidelines for Book Longevity of the Council on Library Resources.

10 9 8 7 6 5 4 3 2 1

THIS VOLUME IS DEDICATED TO

Canace, Sawyer, Lily, and Oren

IN THE HOPE THAT THEY GROW UP IN A WORLD

WITH RESPECT FOR RIVERS

Contents

Preface

The term *virtual reality* refers to the use of computers to generate an image or environment that appears real to the senses. Such a virtual environment can serve many useful purposes, whether training surgeons or airline pilots or simply providing amusing video games. But a virtual environment of course cannot exist outside the computer; it is a human construct and cannot function on its own. The antithesis of a virtual environment is a healthy natural ecosystem which is fully functional in the absence of humans. There are no ecosystems remaining in the twenty-first century that do not experience some impact from humans, but the degree varies widely. Some ecosystems are obviously heavily affected. The eastern margin of the Front Range in northern Colorado, for example, has an urban corridor of more than three million people and is surrounded by extensive croplands and animal feedlots. The lower South Platte River flows through this landscape. Numerous scholarly papers describe the process of river metamorphosis which occurred along the lower South Platte during the nineteenth and twentieth centuries, causing the channel to become more narrow, sinuous, and surrounded by streamside forests. These channel changes have altered the habitat of native warm-water fish species and migratory birds such as the endangered sandhill crane. Concern over these alterations, and over potential problems with water quality and quantity, has led to regular public forums in which technical experts and concerned citizens meet to discuss how best to manage the natural resources of the lower South Platte River corridor.

In contrast to the lower South Platte River, the mountainous upper South Platte basin is more likely to be viewed as an almost pristine region; relatively few people are aware of how nineteenth- and twentieth-century land-use patterns impacted the mountain rivers of the South Platte basin.

When I moved to Colorado in 1989, I was impressed by the sparkling water of the mountain rivers, and I too assumed that these were natural, fully functional rivers. It was only after I began to read historical accounts of the Colorado Front Range and to examine the streams more closely that I realized how dramatically they had been altered. I began to think of them as *virtual rivers,* which had the appearance of natural rivers but had lost much of a natural river's ecosystem functions. Out of this realization grew a desire to write this book in the hope that it would enhance the reader's awareness and understanding of how human land-use patterns may have altered seemingly pristine streams. If people are to make wise and informed decisions about the use of natural resources and the management of rivers in the Colorado Front Range in the future, they must first understand the nature and scope of the changes that have already occurred.

Although *Virtual Rivers* focuses on the upper South Platte River basin, the interactions between human land-use and river systems that are described in this book are applicable to many mountainous regions worldwide in both developed and developing nations. These areas are coming under increasing pressure as world population continues to grow and as issues such as hydroelectric power generation are pushed forward in developing countries. Recognition of this pressure can be found in the establishment of journals such as *Mountain Research and Development,* e-mail bulletin boards such as Mountain Network, and nonprofit groups such as Woodlands Mountain Institute, all of which are devoted to scientifically sound and sustainable development of mountainous regions. It is in this climate of growing concern over mountain resources and ecosystems that I think *Virtual Rivers* will find a useful place.

Virtual Rivers is designed to serve as a reference for both the specialist and nonspecialist. As a specialist reference, the book includes extensive citations of technical literature. As a book for general readers, *Virtual Rivers* is written in nontechnical language, and bibliographic sources are cited as notes rather than in the standard scientific format. The book contains abundant illustrations to help explain technical concepts and to help the reader visualize the land uses described. The text is organized chronologically, beginning with beaver trapping, the earliest Euro-American activity that heavily impacted rivers in the Colorado Front Range, and continuing through to urbanization and other effects of the early twenty-first century. As with any other form of history, only by understanding where we have been can we understand where we are going.

Acknowledgments

Many individuals have provided me with assistance while I was research-ing and writing this book. I would particularly like to acknowledge the efforts of the following people. Clifford Blizard did an excellent job of identifying and summarizing many of the original historical sources used in the book, and he also made many helpful comments on the first draft of the manuscript. Marsha Hilmes and Annette Wohl also made useful suggestions for a later draft of the manuscript. Robert Behnke, Andrew Miller, Cathy Tate, and several anonymous reviewers helped to make the manuscript more comprehensive and accessible to the gen-eral reader. Jean Thomson Black of Yale University Press believed in this project, and her encouragement and persistence were vital in trans-forming the manuscript into a book. Reference staff were very helpful in obtaining access to historical photographs at the Fort Collins Public Library, the Denver Public Library, the Colorado Historical Society, the American Heritage Center of the University of Wyoming, and the Car-negie Branch Library for Local History of the Boulder Historical Society. The Colorado Water Resources Research Institute provided financial as-sistance in purchasing copies of historical photographs for use in this book. Finally, I would like to thank my colleagues and students at Colo-rado State University. They have helped me, through numerous stimu-lating discussions, to understand both the form and function of rivers.

Virtual Rivers

Chapter 1

Of Rivers and Virtual Rivers

> July 29, 1843. . . . It was a mountain valley of the narrowest
> kind—almost a chasm; and the scenery very wild and beautiful.
> Towering mountains rose round about; their sides sometimes
> dark with forests of pine, and sometimes with lofty precipices,
> washed by the river; while below, as if they indemnified them-
> selves in luxuriance for the scanty space, the green river bottom
> was covered with a wilderness of flowers, their tall spikes some-
> times rising above our heads as we rode among them.[1]

John Charles Frémont of the Topographical Engineers was one of the
first people to write of the mountain valleys forming the upper drainage
of the South Platte River. On the way to the Columbia River and California in 1843, Frémont and his expedition followed the valley of the Cache
la Poudre River into the Front Range of the Colorado Rocky Mountains.
This was the region that enthusiastic travelers would soon be calling
the Switzerland of America,[2] but in the 1840s the area was essentially
unknown to Euro-Americans. Within the next three decades that situation changed swiftly, as discoveries of gold and silver in 1859 and the
westward movement of railroads and agriculture brought thousands of
people to the mountains.

Those people who would change the landscape of the Colorado Front
Range first approached the region from the east in their canvas-covered
wagons, gradually leaving behind them the settled and prosperous farmlands of the States. As they traveled westward, the woodlands grew thin-
ner and more scattered, giving way to the rolling grasslands of Illinois
and Iowa. Most of the pioneers began their journey in spring, when the
prairie rivers flowed broad and muddy with snowmelt, and the grass was
just greening. The land opened out around them, the trees confined to
the watercourses in lines of darker green cutting across the land's con-
tours, and the horizons blurred at the distant meeting of grass and sky.

The upper canyon of the Cache la Poudre River.

They named this land the Great American Desert. Ahead of them stood mountains so high that snow was said to last all summer, and beyond the mountains desert again. As they followed the Platte River, the pioneers must have squinted into the late afternoon sunlight toward the western horizon, wondering whether the dim white shapes they saw there could be mountains or just summer thunderheads. Then one morning the pioneers would see a dark, undulating band low on the horizon, its top just brushed with white, and perhaps a shout might go up along the wagon train. As they grew closer the dark band rose and became more substantial, until it must have seemed they were facing an impenetrable barrier.

Driving west across the Plains today on any highway, you can watch this same progression from mirage to reality as the mountains solidify

The homestead of Tom and Mollie Morgan in Poudre Canyon, 1887.
(Photograph courtesy of the Ft. Collins Public Library)

on the horizon. On heat-soaked days of midsummer, much like those
the pioneers must have experienced, the air may be thick with humidity
so that the Plains, the base of the mountains, and the sky all blend into
a vague aquamarine distance. But the elevational band that is still under
snow stands out from this background in a striking white slash, a white
so intense that it glows. The highest peaks form a bold line beginning
north from Longs Peak and continuing south to near Boulder, to a gap
of lower, forest-green mountains where Clear Creek and the forks of the
South Platte River flow onto the Plains, and then the white massif of
Pike's Peak on the horizon south of Denver. This white band is winter's
treasure, the snow and ice that will melt slowly through the summer,
nourishing all the life that depends on the hundreds of channels intri-
cately cut into the resistant rocks of the Front Range.

The people who settled along these channels beginning in the 1860s
changed the landscape, cutting trees, building roads and houses, and di-

verting water from the stream channels to meet their needs. Accounts published by early travelers to the mining regions contrast vividly with Frémont's description of Poudre Canyon. Traveling along Clear Creek from Golden to Black Hawk in 1866, James Meline wrote:

The gulch miner has been here in all his pristine strength and glory. Gravel, sand, bowlders, rocks — not one stone left upon another; not one where Nature put it. The entire bed of the stream in the condition of the Kentuckian who was "uneasy in his mind." It was all "tore up." Here sent as high up the bank as impossible hydraulics would allow, and left to feel and trickle its way along a steep mountain side some hundred feet below, that crowbar, pickaxe, and spade might hold high revel in its quiet bed; there put into the straight-jacket of a race, to feed that water-wheel now rotting in the dam above it. Old cradles, broken rockers, quartz pounders half completed and abandoned. . . .[3]

Once mining ceased in the early 1900s, vegetation began to grow back along the valley bottoms and logged slopes. New people coming to the Front Range had no idea of the region's appearance prior to 1859, and as the changes caused by activities in the 1860s and 1870s became less obvious, people largely forgot that such activities had ever affected the landscape.

More than a century later, people continue to change the channels and hillslopes of the Front Range, often without realizing how much of this landscape may already have been altered by human activities. Hikers along the trails of the Front Range commonly encounter old roadbeds slowly going back to forest, or an abandoned pile of mine tailings standing out bright orange against the green slopes. But it is difficult to mentally integrate the effects of all of these activities and to imagine the landscape as it appeared prior to the coming of people of European descent. Thus the early records of explorers such as Frémont, which provide our first glimpse of the upper South Platte River basin, also provide us with a baseline against which to assess change in the intervening 150 years.

The upper drainage of the South Platte River covers an area approximately 160 miles (275 km) south from the Poudre River to the South Fork of the South Platte, and 60 miles (100 km) east from the Continental Divide to the base of the Front Range. It is a region rich in contrasts, from the alpine tundra 14,000 feet (4340 m) above sea level along the spine of the mountains, to the semiarid grasslands at 5,000 feet (1700 m) elevation. Climate varies with altitude, from an annual average of 40 inches (100 cm) of precipitation and 36 °F (2 °C) at the highest moun-

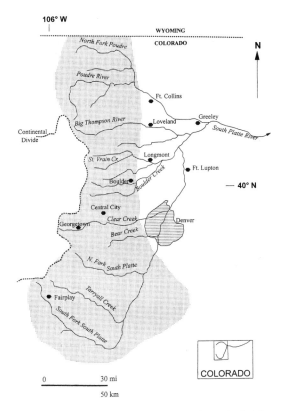

106° W

WYOMING
COLORADO

N

North Fork Poudre

Poudre River

Ft. Collins

Big Thompson River
Loveland
Greeley
South Platte River

Continental
Divide

St. Vrain Cr.
Longmont
Ft. Lupton

Boulder
Boulder Creek

— 40° N

Central City
Clear Creek
Denver
Georgetown
Bear Creek

N. Fork South Platte

Fairplay
Tarryall Creek

South Fork South Platte

0   30 mi
          50 km

COLORADO

Index map of the Colorado Front Range, here shaded from the Continental Divide east
to the base of the mountains. Principal towns and drainages are named. Inset shows
location of Front Range in State of Colorado.

tains, to 15 inches (40 cm) and 50 °F (10 °C) at the mountain front.[4]
Climate in turn controls soil formation and plant communities, and the
Front Range has distinct elevational bands of vegetation: alpine tundra
of ground-hugging grasses, cushion plants and dwarf trees at elevations
of 14,000 to 11,500 feet (4300 to 3480 m); subalpine forests of spruce
and fir down to 9,000 feet (2800 m); and montane forests of mixed conifers, aspen and other deciduous trees down to the transition to steppe
vegetation at about 5,000 feet (1700 m).[5] Through these disparate zones
flow the rivers and creeks.

Rivers are the lifelines of settlement in the arid western United
States. Their valleys have been lines of travel, from the seasonal migration routes of hunter-gatherers of the past to today's east-west interstate
highways. Moving water has a strong esthetic appeal. People have settled

The upper drainage basin of the Big Thompson River, late June 1995. The highest peaks have alpine tundra, while the valley bottom is covered by subalpine forest.

Semiarid steppe vegetation (grasses and shrubs) at the base of the Colorado Front Range, near the mouth of Poudre Canyon.

along rivers for their beauty as well as for immediate access to water for agriculture and industry, all too often expecting the rivers to carry their waste or mistakes downstream and out of sight. A river ultimately absorbs everything that happens in its drainage basin, and the basin is the fundamental organizational unit for the physical, chemical, and ecological flow of materials and energy. Rivers thus reflect the cumulative historical effects of our activities.

I moved from the Arizona desert to Colorado in 1989. Rivers in the desert of central and southern Arizona no longer flow except after rainfall, although they once flowed steadily, supporting beaver colonies and dense stands of mesquite and cottonwood. My first impression upon reaching Colorado was that the cold, clear waters I saw flowing along streams in the Front Range reflected a largely untouched mountain range. I was surprised when, two years later, I read Frémont's description of the Poudre River. I thought that I had come to know the Poudre well in the course of doing field research, but I had never seen the beaver dams, lush streamside forest, and other features that Frémont described. Intrigued by the discrepancies, I began to search out other historical records of Front Range rivers. As I examined nineteenth-century photographs of hillslopes bared by timber harvest and stream channels excavated for gold, I came to realize the extent of human alterations to the Front Range's plant and animal communities, to the chemical and physical processes occurring on hillslopes and in rivers, and ultimately to the landscape as a functioning ecosystem. As I thought about the increasing human population of the Front Range, and the decisions that are now being made regarding protection and alteration of rivers, I realized that other people also needed this historical perspective.

The purpose of this book is to examine how land-use changes during the past 150 years have affected the streams of the Colorado Front Range, and, by extension, mountain rivers throughout the world. Although the focus is on the Colorado Front Range, I have included numerous examples from mountain rivers elsewhere. Human activities and the resulting changes in these rivers have some basic aspects that are common to mountainous regions everywhere, and these are emphasized in this book. Studies of river channel changes resulting from mining, for example, have indicated decreased channel stability and downstream deposition of sediment along mountain rivers in Colorado; the

Bear, American, Yuba, Sacramento, and other rivers in California; and mountain rivers in Alaska, Tasmania, Colombia, and Malaysia. It is these commonalities that allow a detailed study of the Colorado Front Range to serve as a representative example of mountain rivers throughout the world.

One of the key considerations when examining human impacts on mountain rivers is that of human perceptions of these rivers. Because we view a given landscape for at most a few decades, we often do not recognize either gradual changes in the landscape that may occur over longer time spans, or the possibility that even a century ago the landscape may have looked very different. The more sparsely settled mountains of the world are particularly likely to give the impression of wild, untouched, and unchanging nature. But the world's mountains, and the rivers that flow through them, may have been as heavily impacted by humans as the dammed, channelized rivers of the lowlands. The important difference is that people are less likely to perceive the mountain rivers as being altered by human activities. And unless we recognize these alterations, we cannot understand or effectively protect or manage mountain rivers.

Human impacts to mountain rivers are accelerating as world population continues to increase, thereby forcing people into agriculturally marginal mountain environments; increasing the demand for mountain resources of timber, water, and minerals; and increasing the numbers of people seeking recreational and esthetic escape in the mountains.[6] These population pressures are exemplified by the rapidly growing urban and suburban centers along the arid eastern margin of the Colorado Front Range, a topic which was the subject of a 1996 article in *National Geographic* magazine.[7] The Front Range provides a recreational and esthetic escape and is a vital water source as well. The rivers of the Front Range are dammed and diverted to provide water to the surrounding regions, and they are constricted by roads and railroads, stocked with fish, and polluted by mining wastes and urban runoff. Yet many of the people who visit the Front Range for short periods perceive the landscape to be a nearly pristine wilderness because they are unaware of the historical impacts of human activities. This ignorance of both history and of the difference between a diverse, stable river ecosystem and a river which has been simplified to a canal carrying water downstream is a problem that may confront people attempting to understand and manage mountain rivers anywhere in the world. By exploring the changes that have occurred along the Front Range rivers, we may come to understand the difference between rivers and virtual rivers, which give the illusion of being fully functional but are in fact no longer rivers in the truest sense.

The Study of Rivers

There are many ways to study a river. Contemporary scientists and engineers tend to divide the study of rivers into many narrowly defined portions, with each investigator delving deeply into a single portion. Civil engineers may focus on sediment transport along the river channel; aquatic ecologists may focus on the community structure of insects; fisheries biologists may be most concerned with the predator-prey interactions among fish. The study of historical changes along the rivers of the Colorado Front Range that is presented in this volume is organized around the framework of river geomorphology: the study ("ology") of earth ("geo") surface forms ("morph") and processes—the science of landscapes. Both the forms of rivers, and the processes that create those forms, come under the province of river geomorphology.

Rivers and drainage basins are shaped by processes occurring across space and time. At the largest scale are the factors that are not affected by processes in the river channel: climate, lithology, tectonics, and basin physiography.[8] Lithology refers to the type of rock underlying a drainage basin, which is an important control on the amount and size of sediment created through weathering. In the Front Range, for example, shale weathers to clays that produce gently curving hillslopes. Granite, on the other hand, weathers to sand and gravel, and often produces hillslopes with steep cliffs. Tectonics are the vertical and horizontal movements of the Earth's surface that may alter channel gradient. Basin physiography refers to the shape of the drainage basin and includes the distribution of landmass within the basin with respect to elevation, the gradients of hillslopes and river channels, and the spatial arrangement of rivers. These variables, together with vegetation, soils, and the patterns in which people use the land surface, control the movement of water and sediment from the hillslopes to the rivers.

At a more immediate level, rivers respond to the discharges of water and sediment entering from the hillslopes and moving along the channel. Any change in these controlling variables will cause a corresponding response in the river. For example, if a dam built across a river traps sedi-

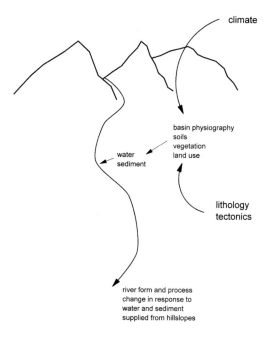

A river in a landscape context.

ment moving downstream, the loss of sediment downstream from the dam may cause erosion of the channel bed and banks. Water flowing in the river has energy available for transporting sediment, and if sediment is not introduced from other sources, it will be mobilized from storage points along the river. The magnitude of river response will be a function of the magnitude, rate, and duration of the external change.

Often there is a lag time before a river responds to a change in water or sediment discharge, and this may be caused by the existence of a threshold. A threshold separates two distinct modes of operation of the river system. For example, a severe forest fire may destroy the vegetative cover that tends to stabilize hillslope sediments. The first heavy rainfall following the forest fire may destabilize the hillslope and cause a landslide that introduces a large amount of sediment into the river. Because flow in the river is unable to transport all of this sediment downstream immediately, over a period of months to years the river may change its geometry from a meandering to a braided pattern with multiple, rapidly shifting channels. This new pattern may persist for decades to centuries until sufficient sediment has been removed and the river once again crosses a threshold and assumes a single channel.

The response of a river to some external change may not be synchronous along the river.[9] Downstream portions of a drainage basin may be affected by changes upstream, and vice versa. It thus becomes important to consider how land-use changes in one portion of a basin may affect the remainder of the basin. Land-use impacts on downstream channels led to the establishment of forest reserves in the Colorado Front Range as early as the 1890s.

In the late 1880s the American public became alarmed at the depletion of natural resources, particularly in the western states and territories. Much of this concern focused on timber, which had been decimated in regions such as the Colorado Front Range, where the rapid growth of settlements associated with mining, and the demand for railroad ties, led to extensive cutting of forests between 1859 and 1890. Photographs from the 1880s and 1980s of the same location dramatically illustrate changes in forest cover, resulting in part from timber harvest prior to the 1880s photographs. There are a great many more trees in the recent photographs, and these are mostly even-aged stands 60 to 100 years old.[10] Farmers and ranchers who had settled in the lower parts of mountain watersheds recognized that timber harvest in the upper basin increased sediment erosion from the hillslopes and silted their irrigation structures. These lower-basin citizens complained to Congress, which passed the Forest Reserve Act in 1891. This act enabled a president to establish reserves of federal land, and in 1902 these reserves were withdrawn from homesteading.[11] Forest reserves such as the Medicine Bow Forest Reserve (1897) in the Colorado Front Range were some of the earliest established.

Human Impacts on Rivers

Human activities that affect rivers can occur either directly within the river channel or outside the channel but still affect the movement of water and sediment from hillslopes into the channel.[12] Several historic activities in the Front Range directly altered the channel boundaries or the movement of water and sediment within those boundaries. Early fur trappers killed large numbers of beaver, reducing the number of beaver dams. These dams regulated the movement of water and sediment through the channel, slowing the passage of flood waves and serving as settling basins to trap sediment. When the beavers were killed, the dams were destroyed, allowing floods to cause channel downcutting and increased sediment transport. As railroads reached the Rocky Mountains and western Great Plains, lumber for railroad ties was often floated down Front Range channels in huge masses consisting of thousands of logs. These log masses acted like giant scouring brushes on the channels, ripping out streamside vegetation, destabilizing sediment stored along the channel bed and banks, and removing channel-bed irregularities such as pools. To remove the gold flakes mixed in with the sand and gravel along channel beds, miners moved substantial portions of the channel-bed sediment, which often resulted in decreased channel stability and increased transport of sediment downstream. Diversion of water for irrigation and mining resulted in decreased sediment movement in some rivers and changed the natural seasonal fluctuations of others. Irrigation reservoirs built along the Front Range channels trapped sediment and caused increased channel erosion where the sediment-free water left the reservoirs. As a result, some channels suffered minor effects, while other channels experienced multiple impacts. Taken together, these activities have substantially modified both channel morphology and the movement of water and sediment along the Front Range rivers.

Similar direct modifications of river channels have occurred in nearly every mountainous region in the world. Beaver were trapped in the mountain rivers of Europe as early as the twelfth century, becoming locally extinct by the sixteenth century.[13] By the seventeenth century, people were busily trapping beaver in eastern North America, moving steadily west and north across the continent during the next two centuries, from the Appalachian and Allegheny Mountains to the Rocky Mountains, the Sierra Nevada, and the Cascade Range on the West Coast, and up into the mountains of northern Canada and Alaska. The modification of river channels to facilitate the passage of railroad ties or other waterborne products has occurred in mountainous regions throughout North America[14] and Europe. Mining of metals mixed in with riverbed sediments has dramatically impacted river channels in regions with precious metals. Well-studied examples include the Rocky Mountains, the Sierra Nevada of California, the Colombian Andes, northeastern Tasmania, and the mountains of interior Alaska.[15] Construction of water diversion and reservoir structures has been particularly widespread in mountainous regions, which often provide a large portion of the water supply to the surrounding lowlands. These types of structures began to be constructed in the 1860s in the Colorado Front Range, but they have a much longer history in some regions. The earliest-known dam was built circa 2800 B.C. in Egypt, but the construction of dams taller than 50 feet (15 m) has accelerated greatly since the 1950s.[16] Today, dams affect rivers in nearly every mountainous region in the world. Other direct channel modifications that are not common in the Colorado Front Range but have affected mountain rivers elsewhere include channelization and the construction of sediment dams and levees to reduce hazards from floods and debris flows.[17] These structures are widespread in the mountains of Japan, the European Alps, and the mountains of China.[18]

Indirect human activities are those that affect the movement of water and sediment from hillslopes into channels. In the Colorado Front Range, the removal of protective slope vegetation during timber harvest caused increased rates of water and sediment movement from hillslopes. Increasing road density and urbanization within the Front Range also caused increasing water and sediment movement from hillslopes, because roads generally have a lower capacity to absorb precipitation than do the surrounding natural surfaces. Roads may also alter hillslope stability and trigger landslides or debris flows that introduce large quantities of sediment into the stream channels. And mining has created tailing piles from which sediment has been eroded into the stream channels.

Timber harvest has affected river channels in every forested mountainous region in the world: the Himalayas, the Andes, the European Alps, the Appalachians, the Rockies, the Sierra Nevada, the Atlas Mountains, the Cascade Range, the Urals, the Ozarks, the Sierra Madre of Mexico, Australia's Great Dividing Range, Peninsular Malaysia, and on and on.[19] Worldwide, the decrease in slope stability and the changes in water and sediment yield associated with timber harvest probably have affected mountain rivers most significantly. Urbanization, and the associated construction of roads and railroads, have been locally important in the Appalachians, the Rockies, the European Alps, the Cascades and other coastal ranges of western North America, the Andes of South America, the Great Dividing Range of Australia, and the numerous mountain ranges of southeast Asia, Indonesia, Japan, and Central America.[20] The impacts of grazing and cropping in mountainous regions are of particular concern at present in the Himalayas, many areas of the Rocky Mountains, the Sierra Nevada, and the mountain ranges of

Africa. From prehistory to the present, whenever people have begun to alter native vegetation by keeping domestic animals or growing crops, the resulting increase in sediment movement from the hillslopes leaves a record in the sediment stored along river valleys. These sedimentary records have been studied along mountain rivers in the European Alps, the mountains of North America, and the Himalayas.[21] Destabilization of slopes in association with hard-rock mining has been documented in mountainous regions of Papua New Guinea, Malaysia, western North and South America, and the eastern United States.[22]

Although similar types of land-use activities have occurred in mountainous regions throughout the world, the timing and intensity of these activities have varied widely as a function of differences in regional histories and economies. Consequently, the timing and magnitude of river channel changes in mountainous regions have also varied widely.

The type of channel changes caused by direct or indirect human activities may also result from such natural changes as the advance or retreat of glaciers on a long timescale or forest fires on a shorter timescale. Forest fires, for example, cause increased water and sediment movement to rivers over a period of months to years. Although human activities may mimic these natural disturbances, there is often a change in the magnitude, frequency, and rate of disturbance caused by humans. Humans have been called "hyperactive agents of erosion,"[23] because we often cause faster and more widespread change than do natural events.

As the population of the Colorado Front Range continues to grow, human activities increasingly affect the stream channels of the region. The study of human history has been justified with the explanation that if we do not know where we have been, we cannot hope to understand where we are going. This idea might also be applied to the management of natural systems. For better or worse, we must attempt to manage ecosystems by recognizing how our activities have affected, and will continue to affect, these ecosystems and make adjustments accordingly to minimize the negative impacts. This type of management is being initiated along the lower reaches of the South Platte River basin.[24]

Several studies have documented historic channel change along the downstream, Great Plains portion of the South Platte River, from the base of the Front Range to the South Platte's junction with the North Platte River in Nebraska. When the first systematic channel surveys of the lower reaches were conducted in the 1860s, the South Platte River had a channel approximately 1 mile (2 km) wide and 1–4 feet (0.3–1.2 m) deep. River flow peaked during the late spring to early summer snowmelt period and then remained low for the rest of the year. Hundreds of small, timbered islands could be found within the broad channel, but timber was relatively sparse along the valley banks.[25] With the development of irrigated agriculture and the associated canals, dams, and storage reservoirs that regulated river flow, the channels changed from an all-or-nothing regime of large spring snowmelt floods with low flows during the remainder of the year, to a regime of moderate flows spread throughout the year. Regional water tables also rose as water that spread across agricultural fields soaked into the ground. These changes in flow allowed streamside vegetation to grow more densely along the channel banks, because the vegetation was less likely to be destroyed by June floods, and there was a more dependable water supply throughout the year.[26] As streamside vegetation increased, it trapped sediment and increased bank stability, causing the channel to become more narrow and sinuous with time. Some reaches of the South Platte River between Greeley and Ft. Morgan are now one-fourth as wide as they were in the late 1800s, others have decreased to one-tenth or one-twentieth of their former width.[27] This has reduced channel and floodplain habitat used by migrating sandhill cranes, the interior least tern (endangered), the whooping crane (endangered), and the piping plover (threatened).[28] As a result, scholarly symposia and interdisciplinary, interagency committees have been organized to examine possible means of mitigating this channel change. However, there has been relatively little consideration given to historical changes that might have occurred along the mountain channels of the upper South Platte River basin because people are less aware of these changes. This recognition of lowland channel changes, and relative ignorance of mountain channel changes, is reflected in the large number of papers and books devoted to lowland channel change and restoration,[29] and the relative paucity of such literature for mountain regions.

The Physical Framework

The manner in which any mountain river responds to changes associated with human land use will be governed by the physical framework of the drainage basin and the channel. This physical framework is set by

Looking upstream along the Poudre River to a riffle (whitewater at center right) and pool (flat water at lower left and center) sequence.

the geology, which controls rock type, structure, and tectonic uplift, and by the climate, which controls the rate and manner in which rock weathers to sediment, and the amount and distribution of water entering the river. Together geology and climate create a template within which adjustment of the river occurs.

Mountain rivers are characterized by dramatic changes in channel form over short distances downstream. Channel form can be described in terms of width, depth, channel-bed regularity, gradient, and the size of the sediments forming the channel bed. Width and depth are usually expressed as a ratio; a high width/depth ratio indicates a wide, shallow river. The regularity of the riverbed topography reflects the presence of bedforms, which are regularly repeated alterations in the riverbed, such as pools and riffles. Pools are relatively deep portions of the river that occur on average at a downstream spacing of 5–7 times the average channel width. Riffles are topographically higher portions of the riverbed. Pools and riffles are common along mountain rivers. Steeper portions of the rivers are likely to have steps and pools where cobbles, boulders, and logs are arranged across the channel in vertical steps with a small plunge

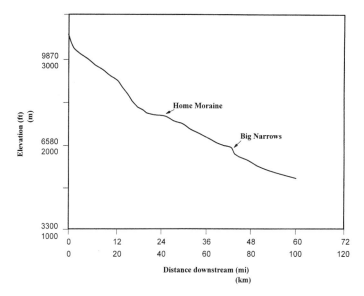

Longitudinal profile of the Cache la Poudre River, Colorado, from the drainage divide to the base of the Front Range. The profile shows change in gradient associated with glacial history and structure. Home Moraine is the furthest down-valley deposit of the Pleistocene valley glacier that occupied Poudre Canyon. This moraine forms a partial dam across the valley and has allowed sediment to accumulate upstream, thus creating a short segment of lower gradient along the valley floor (see photograph of channel reach above Home Moraine). The Big Narrows are an especially deep, narrow segment of the canyon where the trend of the river leaves a large shear zone (see photograph of channel reach in Big Narrows). Elsewhere this shear zone, which is an approximately half-mile-wide (1 km) zone of fractured and faulted rock, is associated with the trend of the valley, and the valley is wider and of lower gradient.

pool at the base of each step. Many of the aspects of river form follow predictable patterns. Steep rivers, for example, tend to have cobble- to-boulder-sized sediments (larger than 25 inches, or 64 mm, in diameter) forming steps and pools. Rivers with lower gradients are more likely to have pools and riffles formed in cobble-to-sand-sized sediment (about 0.3 to 25 inches, or 1 to 64 mm, in diameter).[30]

The downstream changes in form of mountain rivers reflect downstream changes in the factors that control channel and valley form. The advance of valley glaciers during the last two million years, the Pleistocene Epoch, carved portions of the Colorado Front Range canyons into broad, deep troughs. When the ice melted and the glaciers retreated, thick deposits of sediment filled these troughs and created broad sur-

Looking downstream along the Poudre River above the Home Moraine. The channel here
has a bed of cobble and gravel, and flows at a gentle gradient (a vertical drop of
approximately 0.004 feet for every foot traveled downstream, or 0.004 feet/feet).

faces with relatively low downstream gradients. In other areas, rivers
follow zones where repeated movement along faults has created swaths
of sheared and broken rock that are readily removed by weathering and
erosion, leaving broad, relatively flat valleys. Where the path of the river
leaves this shear zone, the canyon becomes narrow and steep. The Big
Thompson River between the Continental Divide and the town of Drake
—a distance of approximately 30 miles (50 km)—begins as a steep, bouldery step-pool channel in its upper reaches, changes to a sinuous pool-
riffle channel in Moraine Park, where glacial deposition has created a
broad, gradual valley, and then alternates between pools and riffles and
steps and pools once it enters Big Thompson Canyon.

At a smaller spatial scale, beavers may also create downstream
changes in river form. Beavers often choose less-steep channel reaches
for their dams, which reduce streamflow velocity upstream and promote
fine sediment deposition, further reducing channel gradient. If beavers
have inhabited a reach for several decades, the river upstream is likely
to be fairly deep and narrow, following a sinuous path through an open
meadow. Here the river may have pools and riffles or a flat, sandy bed.

In contrast to the channel reach above Home Moraine, the Poudre River in the Big Narrows flows over boulders 3–6 feet (1–2 m) in diameter, and forms a series of small waterfalls (average gradient is 0.016 feet/feet).

In contrast, the river segment downstream from the beaver dams may be steep and rocky enough to have steps and pools.

With the exception of localized areas such as those partially dammed by beavers or glacial moraines, the Front Range rivers tend to be steep and rocky, with steps and pools that gradually change to pools and riffles in the lower reaches. Most of the rivers are narrowly confined by bedrock valley walls and are dominated by the cobbles and boulders introduced to the river by rockfall, landslides, and debris flows from the valley walls. These coarse sediments are effectively remobilized only by large floods.

Floods along the Front Range rivers occur either as snowmelt or rain-

In the very gradual valleys (approximately 0.001 feet/feet) created by deposition
immediately upstream from glacial moraines, the Front Range rivers have sinuous, sand-
and gravel-bed channels.

fall floods. Snowmelt floods occur mostly in late May and June, and they
are characterized by peak flows that may last for two or three weeks,
but they seldom exceed a unit discharge of 100 ft³/s/mi² (1 m³/s/km²),
or 100 cubic feet of water per second per square mile of contributing
drainage area. Rainfall floods occur mainly in July and August, when
summer thunderstorms over the Front Range produce intense, localized rainfalls. Rainfall floods are short-lived, lasting at most three or four
days, but unit discharge may reach 3,700 ft³/s/mi² (40 m³/s/km²), or six
times that of a snowmelt flood. These flash floods occur mainly below
7,600 feet (2300 m) in elevation[31] and are particularly effective at transporting coarse sediments.

The floods are superimposed on an annual seasonal cycle of flow in
the Front Range rivers. The highest flows occur from May to July, when
meltwater from the winter snowpack fills the stream channels and summer thunderstorms create intense rainfall over the river basin. Flow then
declines steadily through the autumn, reaching a yearly low from January to March. The magnitude and duration of the annual high flows may
be particularly important in controlling the configuration of the chan-

Downstream from Moraine Park, the Big Thompson River again has a stepped gradient and coarse substrate. (To the right is riprap along a road embankment.)

nel and the amount of gravel-to-clay-sized sediment transported downstream. The annual high flows may also control the timing of the life cycles of insects and fish living in the river. The characteristics of the annual flow regime are thus a key component of the river ecosystem.

Although annual high flows may transport large volumes of finer sediment, floods may cause substantial channel change and sediment movement because they are able to overcome the inertial resistance of the channel boundaries of Front Range rivers. Because these rivers have abundant coarse sediment, they tend to be armored. Armor refers to a surficial layer of coarse sediment on the riverbed that may form through selective removal of the fine sediments from a bed with a wide range of

A pool-riffle reach along the Big Thompson River upstream from Moraine Park in Rocky Mountain National Park. The channel has a bed of cobbles and boulders, and a slope of approximately 0.01 feet/feet.

Aerial view of portions of Colorado Route 34 destroyed during the July 1976 Big Thompson River flood. This photograph demonstrates the erosion common along steep, narrow reaches of the canyon. (Photograph courtesy of Stanley A. Schumm)

Aerial view of highway and houses following the July 1976 flood along Big Thompson Canyon. The cars at the right of the photograph are swirled in the pattern of an eddy. (Photograph courtesy of Stanley A. Schumm)

grain sizes. For example, a debris flow enters a river and deposits everything from clay to large boulders. Over a period of weeks to months, flow in the channel removes all of the smaller sediment exposed at the surface of the riverbed, gradually creating a coarse surface layer of tightly packed cobbles and boulders. These coarse sediments are very resistant to movement, and they protect the finer sediments beneath. But if a flood is powerful enough to disrupt the coarse surface layer, then the smaller sediment underneath is also exposed, and huge quantities of sediment may be transported downstream.

Fine sediment may move along a river suspended in the water column. Coarser sediment moves in contact with the channel bed by sliding, rolling, or bouncing and is called bedload. The exceptionally clear

The lower portion of Bear Creek canyon following a flood in October 1896.
(Photograph courtesy of the Denver Public Library, Western History Department)

waters of the Front Range rivers during periods of normal flow indicate the relatively low suspended loads carried by these rivers. The Front Range consists primarily of crystalline igneous and metamorphic rocks that break down into sediment that is sand-sized or larger. As a result, much of the sediment moved by the Front Range rivers occurs as bedload.

The large floods that transport a substantial portion of the sediment in the Front Range channels occur infrequently on a human timescale. The most spectacular recent example was the July 1976 flood in Big Thompson Canyon. This was a catastrophic event in terms of loss of life and property damage, but floods of this magnitude have occurred repeatedly during the evolution of the Front Range canyons. Considering only the last century, for example, damage to dams, bridges, and roads occurred from floods on the Big Thompson River in 1864, 1894, 1906, 1919, 1951, 1976, and 1982.[32] Similarly, the Poudre River had extreme floods in 1864, 1891, 1904, and 1930. The patterns of channel change associated with these floods are fairly predictable. During the 1976 Big Thompson River flood, steeper channel reaches were scoured of sediment, especially on the outside of bends and in narrow areas. Deposition of sediment occurred on the floodplain in more gradual reaches, and boulders more than 6 feet (2 m) in diameter were deposited on boulder bars in the channel.[33]

Ultimately, the forms of the Front Range rivers reflect the adjustments between water and sediment entering the rivers. Any land-use activity that alters water or sediment input is likely to cause a corresponding change in river form. The response of any portion of the river to a change in water or sediment supply will partly reflect the initial river form. The steeper, cobble-to-boulder reaches of steps and pools are relatively insensitive to changes in water and sediment supply. In contrast, the bedforms of pool and riffle reaches are likely to respond to any change in water and sediment supply.[34]

Pools and riffles are created and maintained by a variety of flow magnitudes.[35] In many rivers, including those of the Front Range, only the largest flows are capable of mobilizing the very coarse sediment that forms the riverbed. During these high flows, velocity and localized stream energy are greatest in the deep pools, and coarse sediment is removed from pools. This sediment may be transported downstream until velocity decreases, usually at the shallower area of a riffle. Pools thus tend to be scoured of sediment at high flows, and riffles tend to accumulate sediment. As the water level drops following a flood, the localized gradient of the water surface becomes steeper over riffles, and some of the sediment deposited at high flow is now eroded and moved downstream. Lower flows may also be important in winnowing fine sediment such as silt and sand from among the cobbles and boulders forming a riffle. This fine sediment is carried to quiet pools, where it forms a veneer over the coarse sediment at the pool bottom.

Pool volume, which is an important control of overwinter habitat availability for fish, is particularly sensitive to land-use activities that change water and sediment movement from hillslopes. If a logging operation destabilized hillslopes enough to increase sediment entering a river, for example, much of the sediment would pass through the steps and pools without substantially altering channel form. Once the sediment reached the pool and riffle reach, however, it would begin to accumulate in the pools, reducing pool volume and creating a more uniform riverbed.

Pool volume is only one example of river form that influences the plants and animals living in or along the river. Water temperature and chemistry, streamflow velocity, the size of riverbed sediments, and many other river characteristics also influence biological communities. For example, the streamflow velocity in which an organism lives affects the energy budget of the organism. A fish resting in a quiet backwater along a channel bank has very different energy requirements than a fish swimming along a steep riffle, where the current flows swiftly. Loss of pools along a river as a result of increased sedimentation will alter the energy requirements of the fish in that river and may ultimately change the species composition of fish able to live in the river. River form is thus crucial to the productivity and biological quality of habitat within the river in the sense that step-pool and pool-riffle reaches may support different organisms.[36] Any alteration in river form will cause a corresponding alteration in biological communities.

Biological Communities

The distribution and abundance of the aquatic plants and animals found in mountain rivers are determined by a complex of physical, chemical, and biological factors, including river form, streamflow velocity, channel-bed composition, temperature, food, competition, and predation.[37] For the Colorado Front Range rivers, these characteristics tend to change fairly predictably along a longitudinal profile from the Continental Divide east to the Great Plains, and the biological communities change accordingly. To use trout along the Poudre River as an example, native greenback cutthroat trout (*Oncorhynchus clarkii stomias*) presently inhabit the highest-elevation tributary streams, nonnative brook trout (*Salvelinus fontinalis*) inhabit the lower reaches of the tributaries, nonnative rainbow trout (*Salmo gairdneri*) are dominant in the middle reaches of the main channel, and nonnative brown trout (*Salmo trutta*) dominate from the canyon mouth out onto the plains. Trout are highly competitive and territorial and, as a result of temperature differences, may be able to out-compete other species only under specific water temperature conditions. Any change in land-use or flow regime that alters water temperatures may thus dramatically alter the distribution of each species of trout in the Front Range rivers.[38]

Native greenback cutthroat trout are physiologically able to inhabit the entire Poudre River basin, but they cannot compete with introduced trout species and are now found primarily upstream from waterfalls that are large enough to restrict upstream movement of the introduced species.[39] Historically, greenback cutthroat trout inhabited all of the Front Range rivers, but the combination of introduced fish species and habitat changes resulting from land use in the mountains stressed the native trout to the point that they are now relatively rare. The history of the greenback cutthroat trout of Twin Lakes, Colorado, provides an illustration of the rapid demise of the species. Prior to 1890 the lakes were noted for abundant greenback cutthroat trout, but during the 1890s rainbow, brook, and lake trout and Atlantic salmon were introduced to the lakes. By 1902 rainbow trout had become dominant, and greenback cutthroat trout were soon extinct in the lakes. Throughout the Front Range, greenback cutthroat trout were declared an endangered species in 1968, but their status was changed to "threatened" in 1978 to facilitate management and restoration. Only five naturally occurring populations of greenback cutthroat trout survived, but artificial introductions established populations in twelve streams and five lakes by 1987, primarily at high-elevation sites within Rocky Mountain National Park.

Trout abundance in steep rivers may be more limited by the availability of physical habitat than by food.[40] Trout require different types of physical habitat at various stages of their life. The cycle begins with spawning habitat, which is primarily gravel-sized substrate under which eggs are buried and allowed to incubate for 60 to 200 days. Spawning habitat is particularly sensitive to changes in the movement of water and sediment: flow fluctuations may expose eggs to freezing or subject them to erosion, fine sediment may suffocate eggs, and extensive deposition may bury them too deeply. As a result, trout egg survival increases with decreasing percentages of gravel finer than about 2 inches (5 mm) in diameter.[41] Once the eggs hatch, the young fish are very vulnerable to predation. To survive, they need nursery or rearing habitat with protective cover and slow flow, which often occur along channel margins beneath logs. As the fish grow larger and stronger they move into adult habitat, which is generally pools or runs at least 1-foot (0.3 m) deep with adjacent faster waters that carry food. Deep pools, log jams, and undercut banks allow the fish to escape bird predators and to rest. Adults also need overwintering habitat of deep water with slow flow and protective cover to survive the cold season, particularly in small headwater streams. Such conditions are typical in deep beaver ponds.

As with changes in water temperature, changes in any of the four primary types of habitat affect trout survival. For example, high suspended sediment loads may blanket redds (nests) with silt, reducing water flow through the redd and thus reducing the flow of dissolved oxygen across the egg surface necessary for embryo development. This effect may be most severe for native species, which spawn during the spring, when sediment loads tend to be greatest. Increased sediment loads that smother the eggs of native trout may favor autumn-spawning nonnative brook and brown trout. An increase in water discharge has been found to correlate with reduced young trout survival because of reduction in nursery habitat.[42]

In total, there are fifteen species of game and nongame fish within the Poudre basin, including trout, western longnose suckers (*Catostomus catostomus griseus*), northern creek chub (*Semotilus atromaculatus atromaculatus*), fathead minnow (*Pimephales promelas*), and longnose dace (*Rhinichthys cataractae*).[43] Increasing fish species diversity in a downstream direction is a general phenomenon in aquatic ecology and likely results from increasing stream size and habitat complexity, for fish "partition" a channel into niches. Streamlined, flattened fish with reduced swim bladders and large fins inhabit riffles, whereas fish with laterally flattened (deep) bodies and large swim bladders occupy pools. Some fish feed on the channel bottom; others are specialized to feed in midwater or at the surface. Some fish are carnivores, some are insectivores or herbivores, others are omnivores. The greater the number of food and habitat types present along a river, the greater the potential for fish species diversity, particularly because young fish often require slower, shallower habitat and smaller food sources than do adults of the same species. Any reduction in habitat diversity associated with land use, such as conversion of pools and riffles to a uniform channel bed as a result of increased sedimentation, will also reduce species diversity.[44]

Other aquatic organisms besides fish may be highly specialized for living in a specific portion of a river. Coarse, stable channel beds such as the cobbles and boulders of steep reaches of the Front Range rivers tend to be colonized by aquatic plants and insects adapted either for attaching and clinging, or for avoiding direct contact with the current. In contrast, depositional areas such as pools are inhabited by organisms that are adapted for sprawling, climbing, or burrowing and that have physiological structures to prevent the clogging of their respiratory surfaces by fine sediments.[45] In addition, many insects spend a portion of their life cycle in the hyporheic zone, meaning the spaces between rocks beneath and beside the surface flow in a channel. This zone may extend for several feet below the channel bed and hundreds of feet across the valley, depending on valley bottom configuration.[46] These aquatic insects are mainly macroinvertebrates, or animals without backbones, that are large enough to be seen with the naked eye—a fraction of an inch (about 0.2 mm) and larger. Macroinvertebrates and aquatic plants provide important food sources for fish, and some aquatic plants such as moss provide habitat for invertebrates. Changes in land use that affect habitat diversity in a stream thus reverberate through its whole complexly interwoven food web.

The majority of organisms in steep mountain rivers are bottom-dwelling species closely associated with the channel bed. As with fish, increasing habitat diversity equates with increasing species diversity, as indicated by richness (total number of species) and evenness (distribution of individuals among the species). In general, macroinvertebrate abundance and species richness are low in the headwater reaches of the Front Range channels, and they increase from the montane zone down to the foothills as a result of increasing water temperature and habitat diversity.[47]

Aquatic macroinvertebrates are often used as indicators of a stream's "health." The distribution of these creatures—mayflies, dragonflies, stoneflies, caddis flies, beetles, and others—is controlled by water temperature, discharge and velocity, channel-bed grain size and stability, suspended matter, and water chemistry (total dissolved solids, dissolved gases such as oxygen and carbon dioxide, and nutrients such as nitrates and phosphates), as well as by aquatic and streamside vegetation. Aquatic invertebrates may thus reflect changes in numerous controlling factors, which may in turn reflect changes in land use within the drainage basin. For example, the highest diversity of bottom-dwelling macroinvertebrates occurs in association with cobble and gravel riffles, where a wide range of sizes in the spaces between the cobble and gravel may be used by different species, and where a channel bed that is stable relative to sandy areas promotes community development. Flooding or land-use activity that destabilizes the channel bed or that introduces fine sediment will reduce species diversity. Mining in the channel bed, road construction, flow regulation, and logging all can cause fine sediments to accumulate in cobble and gravel channels because of an increase in fine sediment availability, resulting in loss of habitat and macroinvertebrate diversity. The time necessary for biotic recovery, or a return to "normal" population levels after a disturbance, will depend on several factors. Species that are more mobile, reproduce more quickly, produce more offspring, and are more flexible in their use of habitat will recover more rapidly. Generally, the recovery occurs via migration of organisms from undisturbed areas. Macroinvertebrates may recolonize a disturbed area through upstream or downstream migration, through vertical migration from within the sediments below the channel-bed, or from aerial sources. The farther downstream a disturbance occurs, the greater the potential for rapid recovery.[48]

River biology may also be considered in terms of riparian corridors. The riparian zone is identified by the presence of streamside vegetation that requires abundant water, or conditions that are more moist than normal. Riparian zones vary in size and vegetative composition as a function of interactions among water distribution and quality, gradient, aspect, topography, elevation, soil, channel-bed sediments, and the plant community.[49] Riparian vegetation along the Front Range rivers varies with elevation in a manner similar to that of hillslope vegetation, which changes from the montane forests upward through subalpine forests to alpine tundra. For individual river reaches, which range from hundreds of feet to miles in length, riparian vegetation also varies as a function of valley-bottom width, soil characteristics, and moisture availability. For example, a gradual reach that had previously supported a beaver dam will have an open meadow with some willow and alder along the channel, even though the channel reach immediately downstream may be steep and narrow, with conifers growing in the sparse, well-drained soil along the channel banks. These differences in valley shape and vegetation also influence land use. Open meadows are more suitable for ranches, whereas narrow channel reaches might support conifers suitable to harvest for railroad ties. Land-use disturbances have thus not been uniform throughout a catchment.

Riparian zones are generally more productive in both plant and animal biomass than are surrounding areas, because the riparian zones have more water and many plant communities and boundaries between plant communities in a comparatively small area, creating habitat for a greater number of species.[50] Both domestic livestock and wildlife of

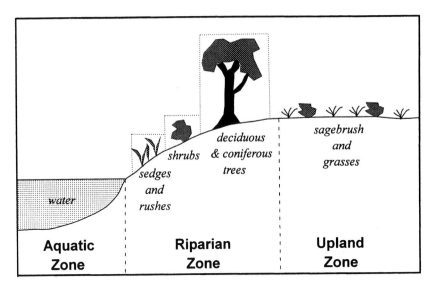

Schematic cross section of a riparian zone showing characteristic plant communities. (After R.J. Smith, 1979, Mountain grazing on riparian ecosystems to benefit wildlife. In O.B. Cope, ed., *Forum — grazing and riparian/stream ecosystems*. Trout Unlimited, Inc., pp. 21–31, Figure 1.)

all types (fish, amphibians, reptiles, mammals, and birds) use riparian zones disproportionately more than they use any other type of habitat. As an example, during surveys of birds in the Colorado Front Range, 82% of all species observed were within the riparian corridors. The riparian microclimate of increased humidity, shade, and air movement attracts mule deer (*Odocoileus hemionus*), and the cover helps the deer to conserve their energy. Riparian zones also provide migration corridors for birds, bats, deer, and elk, among other organisms; deer and elk use these zones as travel corridors between high-elevation summer range and low-elevation winter range. Along many rivers, riparian vegetation produces most of the detritus that provides up to 90% of the organic matter necessary to support headwater stream communities. This vegetation also increases the resistance to water flowing overbank, causing lower flood peaks and greater bank stability than occur on comparable channels without riparian vegetation. Finally, a five-year study of small mountain channels demonstrated that logs falling into the channel from the riparian corridor served to dissipate the energy of flowing water by increasing boundary roughness, thus decreasing bedload sediment movement and maintaining channel stability.

The riparian corridor along the Fall River (a tributary of the Big Thompson River) in Rocky Mountain National Park. Vegetational edges occur at the boundaries between meadow grasses and willow shrubs (foreground and center), and between willow shrubs and coniferous forest (rear-center).

Elk grazing in the riparian corridor along the Fall River in Rocky Mountain National Park.

The downstream distribution of all types of stream biota is often explained in terms of the River Continuum Concept,[51] which can be visualized by following a mountain channel downstream. In its upper reaches, the narrow channel flowing among boulders and logs is largely shaded by a dense canopy of riparian vegetation. Because limited sunlight reaches the channel, there is limited algal growth along the channel substrate. The water remains cold throughout the year, and the aquatic

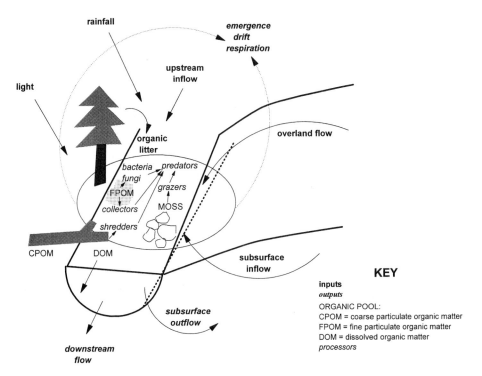

Schematic of the components of a riparian ecosystem. Inputs of sunlight, precipitation, organic litter, streamflow, and surface and subsurface flow from adjacent hillslopes contribute to an organic pool of DOM, FPOM, CPOM (branches, leaves), and diatoms and mosses. Various types of aquatic insects process the components of the organic pool, and outputs take the form of downstream flow, subsurface outflow and, for aquatic insects, emergence (a portion of the life cycle outside of the river), drift (downstream movement), and respiration. (In *Fundamentals of aquatic ecology*, 2nd ed., Blackwell Scientific Publications, Oxford, pp. 230–242, Figure 12.1; I. Valiela, 1991, Ecology of water columns. In *Fundamentals of aquatic ecology*, pp. 29–56).

invertebrates living in the channel must feed on organic debris in the form of leaves and branches that fall into the river. Among these creatures, shredders, as their name implies, can shred this coarse material into smaller pieces. Collectors obtain very small organic debris that moves in suspension within the water column or gathers in quiet areas along the riverbed. Close examination of the riverbed would reveal the filter feeders clinging to a cobble, their fine appendages spread into the current to catch tiny bits of detritus, and the gatherers moving systematically through the organic muck beneath a small eddy.

Continuing downstream, the river grows wider and much of it may be exposed to sunlight. Water temperature changes throughout the year and also fluctuates between pools and riffles. Flow volume may also change as snowmelt increases or decreases, or when a summer thunderstorm drenches the river basin. Smaller cobbles compose the channel bed, and these are likely to be covered by algae. Aquatic invertebrates called scrapers, which feed on this algae, now replace shredders and collectors.

The lower river reaches, which are predominantly east of the mountain front in the case of the Front Range rivers, have reduced water clarity, greater depth, and shifting substrate. These reaches have abundant fine organic matter and are inhabited primarily by collectors.

The River Continuum Concept was developed for eastern deciduous forest streams. This concept may not be fully applicable to the Front Range rivers, which tend to be more open to sunlight in the headwaters, particularly above timberline, and thus to have more algal growth.[52] The concept nevertheless does provide a useful conceptual framework for Front Range rivers, particularly if it is remembered that, as with river form, river biota may fluctuate dramatically over downstream distances of hundreds of feet.

An alternative organizing framework is the concept of process domains.[53] This concept is based on the recognition that spatial variability in geomorphic processes governs the temporal patterns of disturbances that influence ecosystem structure and dynamics. The headwaters of a river, for example, may be shaped by infrequent debris flows, whereas the channel shape and grain sizes of the downstream river reaches may be controlled by more frequent floods. The emphasis in this approach is on reach-scale geomorphic processes and the resulting physical habitat and biological communities, rather than on the downstream trends described in the River Continuum Concept.

Virtual Rivers

The diverse characteristics of the Front Range rivers may all be considered as components of river form and function. Form refers to the physical characteristics of a river—the width/depth ratio and bedforms, for example. Function refers both to physical processes such as the movement of water and sediment, and to the aquatic flora and fauna that have adapted to the constraints imposed by physical form and function.

An example of diversity of form and function along a mountain river. In this downstream view, which includes an approximately 30-foot (10 m) length of channel, the following elements are present: cobble and gravel riffle (foreground); a step in the channel bed, creating plunging flow and small pool (center); deep, quiet water pool with overhead cover (right rear); pool with slightly overhanging bank (center rear); and channel bed material ranging from large boulders (center) to sand and silt (right rear).

Portions of mountain rivers largely unaffected by human activities tend to have high diversity of both form and function: channel gradient and bed materials change over short distances in response to local geologic controls, and this variability in form creates many specialized habitats that can be occupied by diverse organisms. Ecologists and conservation biologists place a high value on diversity, for a diverse ecosystem may be more able to absorb external disturbances. A flood or a landslide, or mining in the channel, may cause habitat changes that eliminate some species. But the greater the diversity of habitats and species in an eco-system, the less the likelihood that any external disturbance will destroy them all.

Relatively pristine portions of mountain rivers are thus stable in both form and function. Stability does not imply lack of change, but rather the ability to absorb change while continuing to remain functional. The effect of various land-use activities in the Colorado Front Range and

The Big Thompson River near the town of Drake. The channel would be steep and turbulent under natural conditions, but channel diversity has been reduced by the construction of straight, riprapped banks, the encroachment of buildings and roads (and the attendant organic and inorganic wastes), and the reduction of natural flow variability.

other mountainous regions during the past two centuries has generally been to reduce the diversity of both river form and function, and thus to reduce channel stability. These simplifications of river form and function may not be readily noticed by an untrained observer. A roadside stream of cool, clear water rippling over multicolored cobbles may be a lovely sight that strikes some observers as evidence of the compatibility of land development and natural habitat preservation. However, a more detailed examination of the stream may reveal water that is contaminated by dissolved chemicals, a fairly uniform channel bottom lacking deep, shady pools, and an annual flow regime that remains constant throughout the seasons because of flow diversion and regulation. All of these characteristics, which are present along portions of the Front Range rivers, mean that the rivers are less able to support riparian and aquatic biota—an impoverishment of river form and function. At some point, function becomes so impaired that we are left with virtual rivers which give the illusion of being fully functional, but are in fact no longer rivers in the truest sense.

As human concern over endangered species increases, we can ask whether it is feasible to save a fish or any other organism without saving the ecosystem in which it evolved. To answer this question we must understand the interactions between fish and rivers, as well as the impact of human activities on these interactions. The remainder of this book explores these issues in greater detail, focusing on the Colorado Front Range. The next two chapters examine the effects of specific land-use activities on mountain rivers, and the last chapter evaluates the present condition of mountain rivers. But first, it is appropriate to review the nature of the Front Range rivers prior to the arrival of people of European ancestry.

Before the Europeans

The rivers of the Colorado Front Range had a long history of adjustment to environmental change before people of European descent arrived in the early nineteenth century. The entire region was repeatedly elevated and worn down in a series of mountain-building episodes lasting tens of millions of years. Masses of glacial ice surged down the mountain valleys and then retreated as global climate cooled and warmed. Forest fires burned across individual valleys, intense summer rains triggered debris flows, and elk grazed the riparian meadows. The rivers were dynamic systems constantly responding to subtle and dramatic changes in the factors controlling the movement of water and sediment from hill-slopes and into the river channels.

The first people to migrate into the region did little to change this situation. There is a lively debate among archeologists as to when humans first settled much of North America.[54] The present consensus is that humans lived in both North and South America well before 12,000 years ago, and there are controversial sites that may indicate human presence up to 32,000 years ago. Archeological sites on the Great Plains of eastern Colorado date to at least 12,000 years ago; those in the mountains seem to go back about 10,500 years. It thus appears that people first migrated into these regions following the most recent retreat of the alpine glaciers. In 1981, there were at least 5,581 known archeological sites on the Colorado plains, but the number of sites in the Front Range was much smaller. This may be partly a function of limited archeological research in the Front Range. Areas that have been examined are gener-

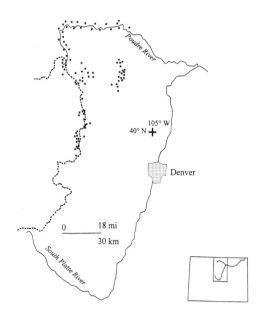

Schematic of known archeological sites in the Colorado Front Range. Sites are not precisely located. The distribution of sites largely represents the distribution of archeological surveys; site density is likely to be equally high in the central and southern portions of the South Platte basin, which have not been examined. (After J.B. Benedict and B.L. Olson, 1978, The Mount Albion complex. Research Report No. 1, Center for Mountain Archeology, Ward, Colorado, Figures 7 and 62; M.P. Grant, 1988, A fluted projectile point from lower Poudre Canyon, Larimer County, Colorado. Southwestern Lore, v. 54, pp. 4–7; J. Slay, Archeologist for the Arapaho-Roosevelt National Forest, personal communication, 10 January 1995; M.E. Yelm, 1933, Archeological survey of Rocky Mountain National Park-Eastern foothill districts. M.A. thesis, University of Denver, 120 pp.)

ally found to be rich in sites. More than 30 sites have been found within a half-mile-wide (1-km) corridor along the Poudre River, for example, and a reconnaissance archeological survey discovered 60 sites in the Rawah Mountains, at the head of the Poudre drainage. A small quartzite quarry that had been used for 8,000 years was discovered in 1994 near timberline at the northern end of the Front Range.[55]

Many of the Great Plains sites correlate with periods of glacial advance in the Front Range, suggesting that people may have migrated between the plains and the mountains in response to climatic variability occurring over hundreds to thousands of years.[56] There is evidence that people converged on the alpine zone of the Indian Peaks region during an episode of warmer climate between 6,000 and 5,500 years ago. Low rock walls on the alpine tundra that are estimated to be 6,000 years old have been interpreted as fences for driving game; at least 42 of these sites have been identified along the crest of the Front Range. Many of these features are associated with the Mount Albion people who occupied the upper South Platte drainage between 6,000 and 5,500 years ago. Sites associated with these people are identified by their distinctive corner-notched projectile points. The population of this group living at the high elevations declined sharply about 5,500 years ago, probably in response to a progressively cooling climate.

These people were probably hunter-gatherers, similar to the people living here when the first Euro-Americans reached the area. As hunter-gatherers, they may have influenced the distribution of plant and animal species through selective harvest and such practices as lighting small forest fires to improve the grazing for herbivores that they hunted. However, these people were almost certainly nomadic, moving to higher altitudes in summer and back to the lowlands in winter, living in fairly small tribal bands, and moving camp whenever the food supply in the immediate vicinity became too low to support them.[57] As a result, they probably did not have substantial effects on landscape processes or river characteristics in the Front Range.

The Ute Indians migrated to Colorado more than a thousand years ago, and they tended to be the dominant group in the mountainous portions of the region up to 1800. Beginning in the seventeenth century, Apaches and Comanches settled in the Great Plains immediately east of the mountains and made periodic forays into the Front Range. In the nineteenth century they were replaced by the Arapahoe and Cheyenne tribes,[58] which bore the brunt of the Euro-American incursions beginning in the early decades of that century. The Plains tribes subsisted mainly on the bison, although they made occasional trips into the Front Range, particularly in summer. None of the early Euro-American explorers in this region mention substantial alterations of the mountain landscape as a result of land-use by these Native Americans. Thus, landscape changes prior to the coming of the Europeans were slow, as when the glaciers retreated during a timespan of 4,000 years, or limited to a fairly small area, as when hunter-gatherers burned a patch of forest to improve grazing for the deer. It was not until Euro-Americans reached the Front Range that dramatic changes began to affect all of the Front Range rivers a few years to decades later. The rapidity and magnitude of these changes led to the extinction of aquatic and riparian species, as river form adjusted to new ratios of water and sediment discharge.

How did the geologic and climatic history of the Colorado Front Range shape the rivers described by the first Euro-American explorers? The repeated mountain-building created a landscape of steep slopes prone to debris flows and landslides. Cobbles and boulders brought into the stream channels by these mass movements were periodically supplemented by boulders carried down the valleys by masses of glacial ice. When the glaciers retreated, their meltwater increased the streamflow and further reworked the coarse river sediments. Today the steep, bouldery rivers have rapid, turbulent flow, but only the largest floods produced by spring snowmelt or a summer thunderstorm are capable of mobilizing the sediments on the channel bed. Some geomorphologists have argued that the general form of the Front Range rivers is relict from the Pleistocene Epoch of 2 million to 10,000 years ago. They hypothesize that the high discharges associated with glacial meltwater largely shaped the rivers, whereas streamflow today is able to move some sediment along the rivers, but not to alter river form substantially. Erosion and deposition associated with glacial movement produced many of the abrupt changes in valley gradient that in turn cause downstream variability in river form and function. The first humans to reach the Colorado Front Range may have subtly affected the Front Range rivers by altering the frequency of forest fires, and thus the movement of sediment from hillslopes into the rivers. But it was not until the coming of the beaver trappers in the early decades of the nineteenth century that the activities of humans began to alter these rivers substantially.

Chapter 2

The Beaver Men, 1811–1859

Beavers have often struck me as the sort of neighbors I might wish for: quiet, industrious, going steadily about their business and assuming that you will go about yours. Many times I have sat on the banks of a lake or a river early in the evening and watched the dark, streamlined head of a swimming beaver trailing ripples across the water. And I have marveled at their tenacity as I encounter thick stumps with the characteristic conically chewed point that a beaver leaves. On land the beavers have lumpy outlines, and they watch my movements suspiciously. But in the water they leave mists of bubbles in their swift wake, bending and twisting seemingly in pure pleasure. It may be difficult then to remember that they are rodents.

The beaver (*Castor canadensis*) is North America's largest rodent.[1] It is an herbivore that subsists on tree bark, aquatic plants, forbs, and grasses; lives in family groups of five to six animals; and builds dams and canals along waterways. It is this last characteristic that makes beavers of particular interest in a history of the Colorado Front Range rivers, for as the density of the beaver population changes, so do channel characteristics. The density is regulated by territories, which are in turn regulated by food supply. On streams with suitable habitat, beaver density averages two to three colonies per half-mile (1 km). Suitable habitat means rivers close to aspen or willow trees, a permanent and relatively constant water flow, valley widths on the order of 150 feet (45 m), and channel gradients less than 15%. Many of the Front Range rivers meet these criteria, and the first Euro-American explorers in the region observed abundant beavers. Describing the foothills region in 1820, Edwin James of the Long Expedition wrote: "In that portion of Defile creek, near which we camped, are numerous dams, thrown across by the beaver, causing

The four-foot-high mound of soil and woody debris in the center is a beaver lodge. The beaver dam originally extended across the river, but has been breached by spring snowmelt floods.

it to appear rather like a succession of ponds than a continued stream. As we ascended farther towards the mountains, we found the works of these animals still more frequent."[2]

Within two decades, Fremont was writing on June 18, 1844 of a mountain tributary of the Platte River: "In the evening [we] encamped on a pretty stream, where there were several beaver dams, and many trees recently cut down by the beaver. We gave to this the name of Beaver Dam creek, as now they are becoming sufficiently rare to distinguish by their name the streams on which they are found. In this mountain they occurred more abundantly than elsewhere in all our journey, in which their vestiges had been scarcely seen."[3]

Yet only a few days later, Fremont writes of meeting a party of six beaver trappers searching for the last of the valuable pelts.

Estimates of the beaver population of North America prior to the coming of the Europeans range from 60 to 400 million, or 6 to 40 animals per half-mile of stream.[4] Beaver occupied nearly all aquatic habitats from the arctic tundra to the deserts of northern Mexico. When a lucrative European market for beaver pelts developed, a veritable army of independent and company-allied trappers set out across western North

Beaver dams have created a stepped appearance along this tributary of the Big Thompson River in Rocky Mountain National Park. The dam across the center is approximately three feet in height.

America in search of the animals. Much of the initial exploration of the region was driven by the fur trade.

The French and Spanish began trapping beaver in eastern North America in the 1600s, with more than 10,000 beaver per year taken in Connecticut and Massachusetts between 1620 and 1630, and 80,000 beaver per year taken from the Hudson River and western New York from 1630 to 1640. Beaver were not intensively trapped in the interior western United States until the Lewis and Clark expedition of 1804–1806 increased awareness of the region's potential. John Colter accompanied Lewis and Clark as a hunter but was honorably discharged from the expedition in 1806 and promptly returned to the Yellowstone River on a trapping venture with William Dixon and Forest Hancock. He was followed by adventurers in the employ of Manuel Lisa, a Spaniard who had settled in St. Louis. Beginning in April 1807, Lisa sent a series of trapping expeditions up the Missouri River watershed and all through the country east of the Rockies. In August 1811 Ezekial Williams, one of Lisa's explorers, started south from Lisa's Fort on the Bighorn River in Montana. He spent a season hunting in what was then known as

Bayou Salade or Salado, today's South Park and the head of the South Platte River, before spending a few months as a captive of the Arapahoe Indians. Although Lieutenant Zebulon Pike explored the southern and central Colorado Rockies from 1805 to 1807, Williams was the first Euro-American explorer to travel down through the mountains from Wyoming, and the first to realize the wealth of beaver in the region.[5]

Auguste Chouteau and Julius De Munn introduced large-scale organized fur-trading to Colorado in August of 1815, when they brought a brigade of 50 men to trap the beavers. In 1824–25 William Ashley led a party of trappers west along the South Platte and Poudre Rivers to the Laramie River and on into Wyoming, and during the early 1830s his successors in the Rocky Mountain Fur Company worked the North, South and Middle Parks of Colorado until the company's collapse in 1835. During 1826–27 James Ohio Pattie and Ewing Young followed a similar route but turned southwest once they crossed the Front Range and followed the Colorado River all the way to southern Arizona.[6]

Within a decade a series of primitive forts sprang up along the South Platte about 40 miles (70 km) east of the mountain front. These log or adobe structures were trading centers that took advantage of their proximity to the beaver-rich mountain rivers, as well as their location along the major north-south trading routes of the Overland Trail, the Cherokee Trail, and the Taos Trail from Ft. Laramie, Wyoming, to Taos, New Mexico. Fort Lupton (1836–46), Fort St. Vrain (1837–46), Fort Vasquez (1837–42), and Fort Jackson (1842–43) were all located within 20 miles (30 km) of one another.[7] Their proximity indicates the intensity of trapping in the 1830s and early 1840s, and their rapid and fairly synchronous demise signals the exhaustion of the resource. It is understandable that Fremont observed few beaver by the time he traveled through the region in 1843–44.

However, the beavers were not completely eradicated, and various historical accounts seem to disagree regarding how rapidly the animals began to recover.[8] At least one historian maintains that beaver were again numerous along the Platte River by the late 1840s. Fort St. Vrain reopened in 1850 but closed within two years. Yet Kit Carson, who had briefly trapped in North Park beginning in 1832, established several trapping posts on the headwaters of the Poudre River during the winter of 1849–50, about the time that trappers from the Hudson Bay Company were leaving the area because of a dearth of beaver. Undoubtedly, com-

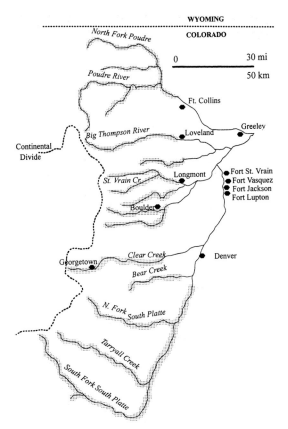

The extent of beaver trapping in the mountain rivers of the upper South Platte basin is shown by the shaded areas. Trapping also occurred in the downstream portion of the river basin.

petition among rival traders, cooperation or the lack thereof by local tribes, and abundance of beaver relative to other regions in the West all played a role in controlling the presence of trappers, so that it is difficult to estimate actual changes in beaver population density. Later travelers to the region seldom mention beavers, their accounts being dominated by descriptions of mining and lumbering activities. Exceptions occur in two accounts from the 1870s. The first is from 1871, when John Tice described old signs of beaver along the Middle Fork of Boulder Creek, writing of beaver dams and of lodges 16 to 20 feet (5–6 m) across and 6 to 8 feet (2–2.5 m) high: "The valley now spreads out to a considerable distance and the bottom land would be as level as an Illinois prairie, were it not for the ridges of the old beaver dams that every fifteen or twenty feet lie across it from one mountain flank to the other and the old beaver lodges. . . . They are even yet from two to five feet high, and four to eight feet wide."[9] Isabella Bird, in an 1873 visit to Estes Park, observed that "the Fall River has had its source completely altered by the operations of the beavers. Their engineering skill is wonderful. In one place they have made a lake by damming up the stream; in another their works have created an island, and they have made several falls."[10]

In the early decades of the twentieth century many of the western states implemented protective laws and began restocking programs to enhance beaver populations.[11] Current population estimates for western North America are 6–12 million animals (pre-European beaver populations for the region are estimated at 40–60 million animals).[12] In 1956 1,000 beaver were estimated to be living in the portion of the Big Thompson River drainage that lies within Rocky Mountain National Park.[13] Today beaver are sporadically trapped along many of the Front Range rivers, but this is primarily for flood-control purposes. Where people have built roads or houses along channels, the water backed up behind a beaver dam may create flood problems.

This history of near-extinction and subsequent protection of beaver also occurred in other regions of the world. The European beaver, *Castor fiber,* was historically found from Britain in the west across the whole Eurasian continent, and from the Mediterranean Sea in the south to the tundras in the north. Hunting and change of land use drove the European beaver close to extinction throughout its range by the nineteenth century; beaver have been extinct in Scotland for 300 years, for example. Sweden subsequently reintroduced beavers in the 1920s, as did other European countries during the twentieth century. Studies of the effects of beavers on river form and function have been conducted in Canada, the U.S. Rocky Mountains, the Adirondack Mountains of New York, the Cascade Range, the Sierra Nevada, and in Europe.[14]

Beavers alter stream channels in a variety of ways. As described by Edwin James, a channel reach occupied by beavers has a stepped appearance. Water ponded behind the beaver dams creates gradual reaches that are punctuated by abrupt drops of 3 to 6 feet (1–2 m) downstream from the dams. Streamflow entering the ponded water behind a beaver dam loses velocity. As velocity decreases, the flow's ability to transport sediment decreases. The beaver pond thus regulates sediment transport and reduces channel bed and bank erosion at and upstream from the pond.[15] Beaver ponds also cause local elevation of both surface and subsurface water levels. The resulting increased availability of subsurface water may enhance growth of streamside vegetation, which in turn fur-

Beaver dams (foreground and right rear) create drops of almost three feet (1 m) in the water surface of this river.

ther reduces flow velocity and traps sediment. As streamside vegetation intercepts contaminants in runoff entering the river from the hillslopes, water quality may be improved.[16] Because beaver dams increase water storage along the river, they promote more uniform stream flows during periods of high and low flow.

As the activities of beavers create areas of ponded water between swifter-flowing reaches of river, aquatic and riparian habitat diversity along the river increase.[17] The greater water depth and aquatic invertebrate productivity associated with beaver ponds provide fish habitat. The ponds and associated streamside vegetation increase waterfowl and animal habitat. Although the beaver dams may provide a barrier to fish migration, individual dams periodically fail or are breached by the beavers when the water gets too high, causing associated changes in macroinvertebrate and fish populations. Through their alteration of rivers, beavers may thus control the habitat available to many other plant and animal species.

As with any other reservoir, a beaver pond acts as a settling basin and sediment trap. With time, the pond will gradually fill with sediment, causing the beaver to abandon that site and move upstream or downstream. When the abandoned dam is breached, the sediment left in the

A series of beaver ponds along the upper reach of the Poudre River in Rocky Mountain National Park.

former pond often supports meadow grasses that may eventually grow back to willow, alder, or cottonwood thickets. The combination of cohesive, fine-grained sediment and lush streamside vegetation produces banks that are resistant to erosion, and abandoned dam sites often have deep, narrow channels meandering through a broad meadow or thicket. If the beaver dam is "prematurely" breached due to removal of the beaver, as must have happened many times during the early decades of the nineteenth century, the stream channel will rapidly cut down through the former pond, producing high sediment loads and dramatic channel widening. In general, flow downstream of beaver ponds contains 50–75% fewer suspended solids than that of equivalent stream reaches without these ponds. When beavers were reestablished along Currant Creek in Wyoming during the 1980s, for example, daily sediment transport decreased from 33 to 4 tons (30,000 to 3,600 kg). Channel gradient decreased, as did bank erosion during spring high flows, which was the main source of sediment to the channel.[18]

The trapping of beavers along the Front Range rivers in the 1830s and 1840s undoubtedly caused changes in river form and function analogous to those described for Currant Creek. These historical changes

A beaver pond surrounded by riparian vegetation (willow, alder, and other plants) in Rocky Mountain National Park.

would have been greatest along the pool-riffle channels preferred by beavers. Pools and riffles provide important habitat for many fish and other aquatic organisms, but these habitats may be altered if the characteristics of water or sediment moving along the river change, as occurs after breaching of a beaver dam. The percentage of the channel bed covered by sand, silt, and clay may increase, for example. Trout prefer spawning habitat that has low velocities along the river bed and clean cobble substrate with few fine sediments. Fine sediments reduce the permeability of gravels, the intragravel water flow, and the availability of dissolved oxygen for developing salmonid embryos.[19] As described in the previous chapter, if too much fine sediment blocks the pore spaces among gravel-sized particles, the trout eggs may suffocate, or the newly hatched fry may not be able to wriggle up into the free-flowing water. Fine sediments can also affect the community structure and abundance of aquatic macroinvertebrates, which are the primary food source for trout.[20]

Trout also do better in rivers where deep pools provide summer rearing and overwinter habitat and refuge from floods. A common channel response to increased sediment, such as sediment eroded from the pond above a breached beaver dam, is preferential filling of the pools. In general, fish and macroinvertebrates have higher species diversity in rivers

Sand and gravel accumulating upstream from a beaver dam (far left) are gradually filling this pond.

where high spatial variability creates a range of habitats that may be used by different species and age groups that have different tolerances for velocity, depth, and cover conditions. Pool filling by sediment creates a more uniform, less habitat-rich channel.[21]

Removal of beaver dams may also change peak flows and coarse sediment transport along a channel, further stressing aquatic organisms. If removal of beaver dams results in shorter duration flood peaks and lower flows between floods, for example, the resulting increase in water temperature may stress aquatic organisms. Enhanced bedload sediment transport, both during floods and lower flows, may kill bottom-living fishes or destroy buried eggs by mechanical grinding and crushing.[22]

Although it was not documented at the time, scenarios similar to those described above may well have occurred along the Front Range rivers in the first decades of the nineteenth century. James described river reaches that resembled a succession of ponds. When the fur trappers removed the beavers that maintained the downstream dams of these ponds, the dams would presumably have been breached or removed by high flows within a year or two. The pools created by the dams would thus have been lost, and the sediment eroded from the ponds would probably have reduced the capacity of any pools down-

stream. Channel bank stability may have decreased as the water table associated with the ponds dropped, killing or stunting some of the streamside vegetation that acts to hold the bank sediment in place. The rivers
may well have undergone repeated episodes of scouring and filling with
sediment before they again stabilized. Although beaver eventually returned to many of these rivers, present beaver population density is estimated to be approximately a tenth of what it was prior to the fur-trapping
era.[23] Similar scenarios can be described for any mountain river from
which beaver have been removed during the past three centuries.

What does the historical removal of beaver mean for the characteristics of Colorado Front Range rivers today relative to the river characteristics in 1800? The rivers today probably have a more uniform gradient,
higher rates of sediment transport, and less-diverse habitat for streamside vegetation, fish, and aquatic invertebrates. Floods, whether caused
by spring snowmelt or summer thunderstorms, probably move through
the rivers more rapidly, transporting more sediment and causing greater
changes in bank configuration through both erosion and deposition.
The rivers have become more efficient conveyors of sediment but less-
diverse ecosystems. However, the channel changes caused by removal of
beaver are probably much less substantial than those caused by changes
in land use that began in the Front Range with wide-scale mining in the
1860s.

Chapter 3

Civilization Comes to the Front Range, 1859–1990

The discovery of gold along the rivers of the Colorado Front Range in 1859 brought a human flood to the region. For every miner who came hoping to dig a fortune from the Earth, there were several other people who came to build roads and railroads, plant crops, raise cattle, and found communities. Many of the activities of these new settlers inevitably focused on the river courses. The rivers held gold buried in their valleys. The rivers carried the water necessary to work that gold and to raft logs down to where railroad tracks were being laid to carry the gold away and to bring in more settlers. In a land of little rain, the rivers provided water vital to raising crops and animals. People took the water as they needed it, assuming their uses of the water had priority over any considerations of the local rivers as functioning ecosystems—such considerations would have been utterly foreign to them. When the demand for water threatened to outpace the supply, the settlers legislated water use and looked for new sources of water beyond the nearest river and beyond the mountains. The use and manipulation of the Front Range rivers during the decades following the discovery of gold altered the rivers' form and function in ways still apparent today, and the patterns of water use that resulted from these activities still govern our decisions about water management and constrain our options for the future.

Gold Fever

There is little mention of gold or other precious metals in the journals of the first trappers and traders to explore Colorado. These men sought furs, pasture, or easy passage across the Continental Divide, and they recorded mainly observations of immediate interest. The mineral wealth of Colorado was not recognized until westward travelers began specifically to look for minerals. In 1849 two separate parties of Georgians bound for California found gold along some of the Front Range rivers, but they were anxious to reach the rich California goldfields and did not act on their findings until 1858.

In April 1858, William Green Russell and several others returned to Cherry Creek, southeast of Denver. Over the next few months they traveled north into the Medicine Bow Mountains and the Laramie River, finding small amounts of gold on many of the rivers. The prospectors occasionally met with other travelers, and one of these, a Mr. Cantrell, took a sackful of gravel back to Kansas City, where he published the find in the newspapers. As James H. Pierce, one of the original expedition members, wrote: "By Christmastime there must have been a thousand men on the South Platte."[1]

The prospectors rushing into the Denver area quickly spread out, prospecting up into Wyoming and down into New Mexico. Clear Creek had the richest placer deposits, and the early mining activity focused there. In May 1859 John Hamilton Gregory discovered the outcrop of a gold lode on North Clear Creek, and in July 1859 the prospectors discovered the rich placer deposits of South Park. Thus, the major gold localities in the upper South Platte basin were all discovered within a few months.

Concentrations of precious metals in the Front Range occur in two forms; as alluvial placers and as bedrock lodes. Alluvial placers consist of pieces of metal ore dispersed through sediments deposited by currents of water or glacial ice. In the Front Range, alluvial placers are found in river-channel beds and banks and in older river and glacial sediments along the valley bottoms.[2] Placers are formed when the chemical and physical alteration of bedrock and the transport of these altered materials remove precious metals from their original location in bedrock, and deliver them to glaciers or rivers. The glaciers or rivers then carry the metals, along with masses of other rock fragments, depositing them in pockets along the way. These pockets often contain sediments ranging in size from silt and clay up to giant boulders, and the challenge for miners is to separate the metals from the rest of the sediment.

Most of the separation methods relied on the fact that gold and many other precious metals are denser than most rock fragments. The simplest separation method, and the first used along the Front Range rivers,

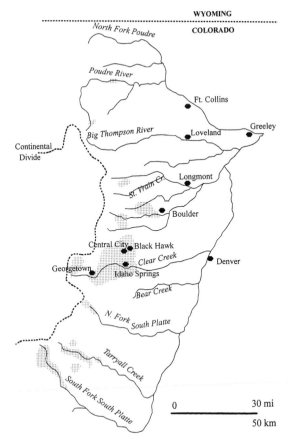

WYOMING

COLORADO

North Fork Poudre

Poudre River

Ft. Collins

Big Thompson River

Loveland

Greeley

Continental Divide

St. Vrain Cr.

Longmont

Boulder

Central City Black Hawk

Clear Creek

Georgetown

Idaho Springs

Denver

Bear Creek

N. Fork South Platte

Tarryall Creek

South Fork South Platte

0   30 mi

50 km

Schematic of the regions of lode and placer mining (shaded areas) in the Colorado Front Range. (After C.W. Henderson, 1926, Mining in Colorado: a history of discovery, development and production. U.S. Geological Survey Professional Paper 138, 263 pp.)

was that of panning. In panning, a miner used a round, shallow metal pan with ridged sides. Scooping up a small load of sediment and water in the pan, the miner swirled it around to concentrate the gold along the ridges or at the bottom. The use of a cradle or a rocker box soon replaced panning. A rocker box is a wooden crate with a sieve at the bottom and a ridged shoot beneath. Sediment and water were dumped into the crate, and then the mixture was rocked. At the largest scale of this type of technology, a sluice—a long, narrow wooden box with washboard-style grooves—was used to separate the gold. Flow was generally diverted from the stream channel along a trough called a flume, and into the sluice, which might be hundreds of feet long. If the water could be pressurized, jets of water could be used to break apart alluvial deposits, which could then be run through a sluice. This was called hydraulicking, or hydraulic mining. Dredge boats were introduced at the end of the 1890s. A

Placer mining along Illinois Gulch, a tributary of Russell Gulch, in the Central City mining region, circa 1860. Water is being introduced from outside the river channel (upper left), which has scoured dramatically relative to upstream reaches. (Photograph courtesy of the Colorado Historical Society)

dredge boat was a floating device containing buckets on a conveyor belt that were used to scoop up channel sediments. The sediments were then run through sluices on the boat, or chemically processed. Methods of chemical processing for placers included amalgamation with mercury, in which the small flakes of ore suspended in water were passed over a surface of liquid mercury to form a mercury alloy, or amalgam. The amalgam was then subjected to fire-refining processes for the recovery of the gold or silver.[3]

The progressive changes in placer mining technology allowed increasingly larger amounts of sediment to be worked for gold. An experienced panner could process approximately 0.5 to 0.8 cubic yards (0.4–0.6 m^3) of sediment in ten hours. Two people operating a rocker box or a hydraulic system, using 100 to 800 gallons (400–3,000 L) of water, could process 3 to 5 cubic yards (2–4 m^3) in ten hours, and a dredge boat could process 7,800 to 8,600 cubic yards (6,000- 6,600 m^3) during the same time interval.[4]

Lode mining took place where the metal was still in place in the bedrock. The primary object in lode mining was to concentrate the metallic ore, which was often disseminated through large volumes of bedrock as small crystal aggregates or narrow, discontinuous veins. Once

Hydraulic mining in the Fairplay district, late 1800s.
(Photograph courtesy of the Colorado Historical Society)

Hydraulic dredge working placer deposits along Clear Creek between 1930 and 1941.
(Photograph courtesy of the Denver Public Library, Western History Department)

chunks of rock containing ore were physically removed from the surrounding rock, they had to be reduced in size. Initially this was accom-
plished with an arrastra, a circular rock-lined pit in which the chunks of
ore and rock were pulverized by stones attached to horizontal poles fas-
tened in a central pillar and dragged around the pit by a draft animal such
as a horse or mule. Arrastras in Colorado goldfields were usually quickly
replaced by stamp mills. Rock in a stamp mill was crushed by descend-
ing pestles, or stamps, that were powered by water or steam. The stamps

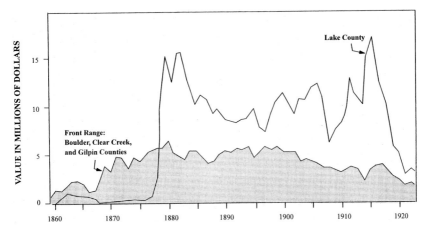

Total value of gold, silver, copper, lead and zinc production between 1859 and 1923 by the principal mining counties of the Front Range, and by Lake County (Leadville) for comparison. (After C.W. Henderson, 1926, Mining in Colorado: a history of discovery, development and production. U.S. Geological Survey Professional Paper 138, 263 pp., Figure 13.)

used in Gilpin County averaged 400 to 600 pounds (200–270 kg) in weight each and dropped 30 times a minute.[5] The pulverized rock was then chemically processed with amalgamation, cyaniding, or chloridizing roasting. Cyaniding involved treating the ores with a weak solution of sodium or potassium cyanide. The solution readily dissolved gold or silver ores from the surrounding rock, and the metals were then precipitated from solution with zinc. When the ores naturally occurred in a sulfide form, the ore could be roasted in cylinders, mixed with sodium chloride to convert the sulfides to chlorides, and then retrieved with pan amalgamation. The mercury, cyanide, and other toxic chemicals associated with mining were often disposed of in a haphazard fashion, by dumping onto the ground or into the nearest stream channel.

Mining in the Front Range has been divided into four stages: (1) placer mining from 1858 to 1860, (2) fissure vein (lode) mining with improving technology, (3) a cyanidation period, with a revolution in ore treatment, and (4) deep lode mining with large capital investments beginning in 1918.[6] Placer and lode mining often overlapped in time and space, and placer mining, which continued into the 1950s, generally involved increasingly larger and more expensive equipment at a given site. Each type of mining affected the Front Range rivers differently.

Table 3.1. A partial chronology of mining activities in the upper South Platte River basin

Date	Location	Activity
Jan. 1859	Gold Run Creek, Boulder County	Gold discovered; 3,000–5,000 people within first season
Jan. 1859 1937–39	Deadwood Digging, South Boulder Creek	Placer mining Dredging
1859 1860	Lefthand Creek and other tribu- taries of Boulder Creek	Placer mining Six-mile (10 km) ditch dug

May 1859
1859–60
1865–1940s
	North Clear Creek; Gregory's
	    Diggings, Blackhawk, Central
	    City
		Gold lode mining and sluicing
		Twelve-mile (20 km) ditch dug
		Forty-mile (70 km) of mining
		    activity; 6,000–7,000 people
		    in Blackhawk and Central City

1859
1860
	Clear Creek	Placer mining
		Hydraulic mining; 60 stamp
		    mills and 30 arrastras in
		    Blackhawk

1861 1864–1913 1934–36	Georgetown	Stamp mill in Idaho Springs Silver mining—mills built Dredging
July 1859–1910 1874–1900s 1919–52	Middle Fork South Platte; Fair- play	Placer mining Hydraulic mining Dredging

1880–84
1887–96
	Poudre River; Teller City, Lulu
	    City, Manhattan
		Silver mining
		Gold mining; stamp mill built in
		    1888

1890s 1902–1930s 1930s	Four Mile Creek	Placer mining and ditches Mill built Dredging operation

The level of mining in the principal mineralized regions of the upper South Platte basin, as judged by total volume of minerals produced, remained between $2.5 and $5 million per year from the late 1860s through the first decade of the twentieth century. This translates into enormous volumes of material being moved and removed. In 1893, for example, 25,838,600 fine ounces (803,580 kg) of silver, 7,695,826 pounds (3,498,103 kg) of copper, 110,000,000 pounds (50,000,000 kg) of lead, 1,650,000 pounds (750,000 kg) of zinc, and $7,527,000 in gold were recovered in the state of Colorado.[7] Assume that anywhere from

Placer mining along Four Mile Creek. (Photograph courtesy of the Carnegie Branch Library for Local History, Boulder Historical Society Collection)

Extensive reworking of the bed of Four Mile Creek in association with placer mining. (Photograph courtesy of the Carnegie Branch Library for Local History, Boulder Historical Society Collection)

1,200 to 2,400 pounds of rock were processed for each pound of metal recovered (1.5–15 kg rock per gram metal), because even the richest ores seldom assayed beyond 20 ounces of metal per ton of rock, and early miners recovered only 60% of assayed gold values.[8] Then imagine this process continuing for seventy years.

A simple north-to-south progression provides an index of levels of mining activity.[9] Relatively little mining occurred within the Poudre River basin, but a gold strike at Manhattan approximately 2.5 miles (4 km) north of the Poudre River led to a short-lived tent city that grew to 10,000 people within a single year. Similarly, silver deposits around the head of the basin supported Teller City and Lulu City from 1880 to 1884. The next major drainages to the south, the Big Thompson River and the upper forks of St. Vrain Creek, were the only major tributaries of the South Platte River that largely escaped the impact of nineteenth century mining activities. The areas immediately south of them were heavily impacted, however. On 16 January 1859, gold was discovered in Gold Run Creek of Boulder County. Three to five thousand people flocked to the site in the first season. Also in January 1859, B. F. Langley found rich placers along South Boulder Creek at a site called Deadwood Diggings because of all the fallen timber in the gulch. Later in the year the creek was flumed in several places to permit its bed to be worked over, and 300 men were at work there. In the summer of 1859, four dams were built where Boulder Creek leaves the Front Range, draining more than a half-mile of the creek bed. There were also placer operations on Lefthand Creek and on the smaller tributaries of Boulder Creek. By the end of the summer of 1860, a ditch had been dug more than 6 miles (10 km) from Lefthand Creek to Gold Run Creek. This permitted water stored in boom dams to be released down the sluices with enough force to separate the gold from the remaining sediment. Similarly, Chinese miners working placers along the length of Four Mile Creek through the late 1890s dug diversion ditches into the creek to increase flow through their sluice boxes. In 1902 the Wall Street Gold Extraction Company built a mill that operated for two years, running tailings directly into Four Mile Creek. The Black Swan Mill, also built in 1902, operated along the lower portion of the creek until the early 1930s. During the depression of the 1930s, a dredging operation that extended a couple of miles above the Wall Street Mill used a steam shovel to scoop up channel sediment, with two sluice boxes to separate the gold. Upstream, a 3-feet-wide (1 m) flume

Placer mining along Gregory Gulch, early 1860s. (Photograph courtesy of the Denver Public Library, Western History Department)

Gregory Gulch in Central City, June 1995.

Gregory Gulch in Central City, June 1995.

diverted the stream so that miners could work the creek. There were over 100 stamp mills built in Boulder County, and extensive placer mining and fluming occurred along Boulder Creek and its tributaries Lefthand, Four Mile, and Gold Run Creeks.

The next major drainage south of Boulder Creek is Clear Creek. Sluicing along North Clear Creek began immediately after John Gregory's May 1859 discovery of a gold lode outcrop on the creek. Within two months, 100 sluices were running at Gregory's Diggings and 500 sluices in the Gregory mining district of Central City and Black Hawk. The 12-mile-long (20 km) Consolidated Ditch was dug in 1859–60 to carry water from the head of Fall River to the Gregory site. By 1865 there was mining activity along 40 miles (70 km) of the Clear Creek drainages,

Central City in 1864, with houses crowding the river channels (lower center and right), and treeless slopes. (Photograph courtesy of the Colorado Historical Society)

and there were 6,000–7,000 people in Black Hawk and Central City. The activity continued well into the twentieth century. In September 1936, for example, 13,590,700 cubic feet (381,760 m³) of gold-bearing gravel were being dug from North Clear Creek daily by two draglines and a power shovel. This daily total would be the equivalent of roughly 42 football fields covered by six feet (2m) of gravel. Flow in the channel was diverted, and the sediment was removed down to the bedrock contact, where gold concentrations are highest. In 1938 Manion Placer Company dredged North Clear Creek into the town of Black Hawk. In 1941 placer gold production in the basin reached its highest yearly value since 1868.[10]

The largest operations occurred along the main fork of Clear Creek.[11] In 1859, there was placer mining along the length of the channel up to the forks of North and South Clear Creek. Hydraulic mining was introduced in 1860. Mining was conducted in river channels and across valley bottoms in the region, particularly near Idaho Springs, where a 20-stamp mill was constructed in 1861. During the summer of 1860 there were 60 stamp mills and 30 arrastras, all waterpowered, operating around Black

Central City in 1995, as seen from Nevada Gulch. Note the return of forest cover. Tourism is high (the parking lot shown here is one of several in the town.) Massive tailing piles are at right and at left rear.

Central City in 1995, as seen from Russell Gulch. Extensive tailing piles remain, but the forest is recovering.

Hawk alone. The region was a veritable beehive of activity, with people working through the sediment in the rivers, building roads along the valleys and up the hillslopes, cutting timber, and building houses, stores, and lumber and stamp mills. Before the gold played out, silver was found near Georgetown in 1864, leading to the construction of many mills for experimentation in silver reduction. The Denver Republican for 6 January 1897 described the construction of two large diverting flumes which took all the water out of Clear Creek for a distance of 3 to 4 miles (5–6 km). Placer mining continued to 1913 between Idaho Springs and Georgetown, but was most active from 1859 to 1863.

The single miner could be quite destructive. When C. M. Clark arrived at Clear Creek in 1860 he noted that the creek was "cold, but exceedingly turbid, as it receives all the 'washings' from the Gregory District, besides those carried on along its own banks and bars." Clark set to work with sluice boxes along a portion of the channel 30-feet (9 m) wide. Because the flow was low, he had to sink the ditch deeper and constructed a wing-dam of large boulders in the stream. The sluice was built of four troughs, each 16 feet (5 m) in length, joined together. With these he worked in a single season a boulder bar 12 feet (3.7 m) above the river that extended back 50 feet (15 m) to the base of the valley wall. The bar was under about three feet of nonplacer sediments, and he dug to bedrock. Clark noted that there were over 200 quartz mills in the mountains, most brought up and assembled in the past season.[12] By the time William Bickham reached Clear Creek in 1879, he noted that individual bars along the channel had been worked a dozen times since 1858.[13]

The other primary concentration of mining occurred along the Middle Fork of the South Platte River,[14] where placer gold was discovered in July 1859, resulting in the founding of the town of Fairplay. Much of this gold was mined from Pleistocene-age glacial deposits. Almost every possible method of placer mining was used along the Middle Fork. Although placer mining continued until 1910, hydraulic mining was introduced in 1874, when two large flumes were built, and continued until the early 1900s. Dredging began in 1919, and portions of the channel were mined until 1952. The Alma placer illustrates the magnitude of mining activity in the region. In September 1883 the Fairplay newspaper reported that a single company had built more than a half-mile (900 m) of sluices. In 1898 the paper reported that another company was working 1,200 acres

(500 ha) of ground to a depth of 50 feet (15 m) with 4 miles (6 km) of ditches from the Platte and two "hydraulic giants." The placer had been worked every summer for 30 years. By 1950 the same placer was still being worked by a portable sluicing plant with a capacity of 1,800 cubic feet (51 m³) per hour.

The changes in the landscape caused by the mining activity of the 1860s–1890s produced abundant commentary from travelers to the Front Range. Traveling through the Central City mining district in 1863, Maurice Morris wrote that Russell Gulch

presents the appearance of a mountain torrent which has suddenly swelled up and brought with it an enormous collection of stones and debris, depositing them on both sides of its normal bed, into which it has as quickly relapsed, maintaining only the complexion of the mountain clays which have defied its purity. On coming nearer, you will observe a number of sluices placed along its channel, and parties of three or four working at intervals of some hundred feet from each other. One of the men is standing by the sluice below the others, throwing away the stones which the water brings down, with a sort of long fork . . . while his confreres dig up the dirt in the neighborhood of the sluice and throw it in, stones and all. This is the whole process of gulch, or surface mining.[15]

Passing through the same area in 1866, James Meline wrote: "The desolation and literal dilapidation of the bed of the creek has now extended up the sides of the valley to their summits. They look like mountains in reduced circumstances, and in a shockingly bad state of repair. Trees and vegetation have long since disappeared. Holes, shafts, and excavations almost obliterate the original surface."[16]

Traveling along the unmined portion of the South Branch of Clear Creek in 1874, Charles Harrington commented that the channel, "unsullied by the refuse of the quartz mills, is a noble stream, rapid and clear."[17] Prior to mining, Clear Creek had abundant trout and beavers, but an 1880 history of the region noted that these creatures were gone. All of the travelers noted that the channels being mined had banks and beds that were thoroughly torn up and destabilized, and waters that were turbid with suspended sediment. Other abuses occurred as well; a photograph of Central City in the mid-1860s shows outhouses built right over Clear Creek.[18]

Most of these descriptions focused on Clear Creek, a principal destination for many travelers. These effects were common to all of the mining regions, however. When a reservoir was built at the mouth of Boulder

Canyon in 1874, Boulder residents could not use the stored water because it was so polluted with tailings from the mines upstream.[19] These effects of mining continued well into the twentieth century. In 1925 ranchers in South Park complained about the decrease in water quality resulting from dredging on the Middle Fork of the South Platte River.[20] The ranchers filed suit against the mining company and halted mining operations. The 1930s dredging operation along Clear and North Clear Creeks discharged tailings directly into the channels, producing high suspended-sediment levels.[21] By that time laws were in place that required companies to impound and store tailings. The testimony of an expert witness for the prosecution in a pollution case associated with mining along Clear Creek in the 1930s noted that water discharged from the mining operations contained up to 16% sediment by weight, making "a thick muddy flow that at times approached the consistency of thin cream."[22]

Mining obviously detracted from the appearance of the Front Range rivers. But did it substantially alter the physical and chemical processes occurring in the rivers and their biota? The effects of mining can be considered under two categories: direct effects that involved mining operations actually carried out in the stream channel, and indirect effects that changed the amounts of water and sediment entering the stream channel from the surrounding slopes. Direct effects of mining included disruption of channel bed and bank sediment, reduction or augmentation of flow in the channel through water diversion, and the introduction of toxic chemicals such as mercury into the channel. Indirect effects included disruption of valley slopes through lode mining and tailings piles, deforestation, and the increase of roads and buildings associated with population growth.

The disruption of channel bed and bank sediment occurred when the sediment was displaced in order to separate gold or other metals dispersed throughout. Travelling along Clear Creek at the height of the placer mining era, Grace Greenwood wrote: "In one place the golden stream [Clear Creek] has been so severely dealt with—its very bed taken out from under it, pits dug beside it, rocks tumbled about—that I exclaimed, 'Surely mining did not do all this; it looks like a convulsion of nature!' 'A convulsion of *human* nature, madam,' said a fellow-traveller."[23]

A 1995 view of the Henderson Mine in the drainage of the West Fork of Clear Creek. The mine tailings and retention ponds are the three massive horizontal features at right-center.

Most of the Front Range rivers have a coarse surface layer that keeps the channel stable and minimizes the transport of sediment except during the largest flows. Once this coarse layer is disrupted, the smaller sediment may be selectively winnowed and carried downstream as suspended sediment that eventually settles in areas of quiet water, such as pools. Suspended sediment generally has an adverse effect on stream biota, as summarized earlier. This is supported by studies of sediment increases associated with placer mining along Alaskan streams.[24] These studies have indicated that mining-related increases in turbidity, total residue concentration, and settleable solids reduced production at the base of the river food web to undetectable levels, primarily by eliminating algal communities. Similarly, the suspended sediment decreased the density and biomass of invertebrates. Increased suspended sediment also affected fish by impairing feeding activity, reducing growth rates, and causing downstream displacement, color changes, decreased resistance to toxins and, in extreme cases, death. Many studies have indicated that excess sediment can smother aquatic plants, abrade or clog respiratory surfaces and collect on the feeding parts of macroinvertebrates,[25] and smother fish eggs and larvae. Under natural conditions, the Front

An aerial view of Clear Creek just east of the mountain front, taken from a balloon on August 13, 1894. The unvegetated bars along the braided channel stand out as lighter than the surrounding farmlands. (Photograph courtesy of the Colorado Historical Society)

Range rivers have relatively low sediment loads during most of the year. The aquatic organisms of most mountain rivers do not thrive in water approaching the consistency of thin cream.

The banks of many Front Range rivers are resistant to erosion because either the bank is composed of very coarse, tightly packed sediment, or the bank is capped by a layer of silt and clay held in place by dense tussock grasses or willow and alder roots. If the resistance of the channel banks to erosion is lowered through the disruption associated with mining, the channel often becomes laterally unstable, repeatedly shifting back and forth across the valley bottom in a braided pattern. An extreme example was described by Helen Hunt Jackson writing of the Platte River just south of Fairplay in an 1898 account: "The Platte River just there is an odd place. It consists of, first, a small creek of water, then a sand-bar, then a pebble tract, then an iron pipe for mining purposes, then another pebble tract, then a wooden sluice-way for mining purposes, then a sand-bar with low aspen trees on it, then a second small stream of water, and lastly a pebble tract—each side of these a frightful precipice." [26]

The rapid lateral movement of braided channels discourages thick accumulations of fine-grained overbank sediment, making it difficult for streamside vegetation to become established. The continual shifting also increases the amount of sediment introduced to the river, causing further increases in suspended load. In addition, the instability of the channel bed is likely to affect aquatic macroinvertebrates by mimicking the effect of frequent large, erosive floods: the erosion and deposition of bed sediments during periods of high flow or channel change can destroy communities or organisms attached to the surface of the channel-bed sediments and cause the downstream displacement of bottom-dwelling animals, which need refuges such as backwaters in order to survive disruptive high flows. In general, streams which flood or are disrupted infrequently may show a characteristic succession of stream flora and fauna with time, in contrast to those that are continually disturbed.[27] The adverse impacts of mining on aquatic organisms have been documented in studies on rivers in France, Canada, and Alaska.[28]

The channel braiding associated with placer mining along the Front Range rivers presents an interesting contrast with historical channel change along the lower South Platte River. In the Front Range, channels that had been confined and laterally stable changed to rapidly shifting, unstable systems, reducing habitat available to aquatic and riparian species that had evolved to inhabit the stable rivers. In the lower South Platte River basin, channels that had been rapidly shifting and unstable changed to confined and laterally stable systems, reducing habitat available to species that had evolved in association with the unstable channels. In both cases, a human-induced change in channel form and function stressed the original biological community but also created new opportunities for species able to survive in the altered environment. The difference between these examples is that colonization of the lower South Platte River by species adapted to stable channels was facilitated by the presence of naturally stable channels farther east on the Great Plains; vegetation, mammals, birds, and fish have all migrated upstream from the east along these river corridors. Recruitment of new species to disturbed rivers in the Front Range has been limited by the absence of naturally unstable rivers supporting species that could migrate to the disturbed rivers. There has thus been a net simplification and impoverishment of the biological communities along disturbed rivers in the Front Range.

Mining-related sediment entering a river may be transported along with the "normal" load in a process termed passive dispersal, or it may so overwhelm the river's transport ability that the channel configuration completely changes, a process referred to as active transformation.[29] One of the earliest conceptual models of mining-sediment transport was developed by G. K. Gilbert in 1917 for the channels of California's Sierra Nevada. Gilbert believed that sediment produced by placer mining in the 1850s was gradually moving downstream as a symmetrical sediment wave, accumulating at a given place in the channel and then being subsequently eroded over a period of about 50 years.[30] Later flume experiments and field studies along channels in the Sierra Nevada and in Tasmania have demonstrated that the downstream movement of sediment is not quite so simple.[31] The initial slug of sediment may gradually spread out downstream, for example, rather than moving as a discrete wave. Also, the sediment often does not move passively down a stable channel. Changes in channel form occur simultaneously, and these may take the form of changes in width-to-depth ratio, the spacing and size of pools and riffles, a planimetric form that changes from meandering to braided, or channel bed slope. Repeated channel alterations may occur in response to a single external perturbation. A century or more may be required before the river returns to a stable condition, or to its premining configuration, as demonstrated by the Middle Fork of the South Platte River.

Historical documents and photographs of the Middle Fork prior to mining, and analogous, unmined rivers nearby, such as the South Fork of the South Platte, provide information about premining channel configuration. The Middle Fork was a relatively deep, narrow stream that meandered through broad meadows of willows and grasses. The channel bed was gravel-to-cobble-size material, and the banks had an upper layer of fine sediment held in place by dense vegetation. Today, reaches of the river that were mined are less sinuous and are often braided. Mined reaches have also been more mobile, as observed in comparisons of aerial photographs from 1938 to 1990. And the mined channel reaches have coarser bed sediments than unmined reaches, although it has been 65 to 80 years since mining occurred in some of these reaches.[32] The mined portions of the Middle Fork flow through the broad intermontane valley of South Park. In contrast, where mined channels are closely constrained by bedrock valley walls, as in many of the Front Range canyons, the channel may respond to increased sediment primarily through changes in bed configuration and gradient, rather than planimetric form.

Among other mountain rivers for which long-term channel changes associated with placer mining have been described are the rivers of California's Sierra Nevada, and rivers in Tasmania, Alaska, and New Guinea.[33] The sediment accumulation and braiding that occurred along these rivers historically is now occurring in areas of active placer mining in the Andes of Colombia.

Reduction or augmentation of channel flow through water diversion associated with mining could also affect sediment transport, channel stability, and stream biota. Reduction of flow would result in decreased ability to transport sediment, which could lead to channel change as sediment coming from the hillslopes accumulates in the channel. Reduction of flow also could stress aquatic organisms through decreased oxygen and nutrients, and increased temperatures. In the extreme cases where flow was completely diverted so that the former channel bed could be thoroughly worked over by the miners, it seems appropriate to consider the river as ceasing to exist for a time.

In contrast, augmented channel flow would increase the river's sediment transport capability. This might be offset by increased sediment introduction from mining, or it might cause erosion of the channel bed and banks. Prolonged high levels of flow during seasons other than the late spring snowmelt could also stress aquatic biota by affecting water temperature and clarity, and the necessity to expend energy. Some fish, for example, do not efficiently process the lactic acid produced by activity. If they are forced to be active for too long—to maintain a position in a strong current, for example—they will die from exhaustion. Most aquatic organisms have adapted to the flow regime of the specific river they inhabit; in the Front Range rivers this includes a long snowmelt peak and shorter thunderstorm peaks at the lower elevations. A study in the Rocky Mountains of Montana, for example, has documented that indigenous organisms are stressed or extirpated in rivers where the flow regime abruptly changes as a result of dams or diversions.[34] Studies of this type were not undertaken on the Front Range rivers in the latter half of the nineteenth century, but the results were undoubtedly similar.

The final direct effect of mining comes from the introduction of pollutants associated with mining. The most general effects of any pollutant are to reduce both community diversity within a channel and the density of rooted aquatic plants.[35] Pollutants may be excess organic matter such as that introduced from outhouses or the excrement of animals, excess sediment, or toxic materials. Degradable organic matter reduces dissolved oxygen and light penetration while increasing turbidity of the water and the abundance of nitrogen in the forms of nitrite, nitrate, and ammonia. (Small amounts of nitrogen are necessary to a stream ecosystem, but too much nitrogen can lead to algal blooms that disrupt the existing food web.) Aquatic insects in the upper reaches of the Front Range rivers, which may constitute over 95% of the bottom-dwelling fauna, are adapted to the cold, clear waters high in dissolved oxygen that normally characterize these streams.

Toxic materials interfere with the respiratory, growth, and reproductive functions of members of the entire stream food web. The toxic materials may act as a time bomb, for they have an impact across time and space. There is the initial introduction, followed by processes of bioaccumulation and biomagnification over a period of years. In biomagnification, some toxic materials are not expelled by organisms, but accumulate in fatty tissues. Any predator thus ingests all of the toxins accumulated by each of its prey organisms, so that concentrations of toxins increase with distance up the food chain, culminating in raptors or humans. Longer-lived organisms may also continually ingest more of the toxin without expelling it, leading to bioaccumulation. In addition, the toxins may be adsorbed onto clay or silt particles, lie buried in a sedimentary deposit, and then be remobilized after hundreds of years by channel-bed erosion or lateral channel shifting during a flood, as documented along channels in England, Spain, Norway, and the United States.[36]

The most common form of toxins in mining regions is heavy metals. Heavy metals may enter a river in solution or be adsorbed to sediment. Metals in solution may come from abandoned tailings piles, which have been found to contain substantial concentrations of heavy metals that may enter the ground and surface water by leaching of precipitation through the pile material.[37]

Most of the metal deposits in Colorado are complex ore, a mixture of copper, lead, zinc, and silver sulfides, and native gold.[38] Pyrite (iron sulfide) associated with this complex ore oxidizes to yield acid water containing high concentrations of iron and sulfate. Oxidation of the other metal sulfides under the acid conditions releases high concentrations of trace elements to the water. Of these, copper and zinc appear to present the greatest danger to resident aquatic life, based on a study correlating presence and abundance of such life with water quality at 995 sampling sites in Colorado. Relative concentrations of these metals may be predicted based on geology. The Central City Mining District is a complex mineral sulfide deposit with well-defined concentric zoning of ore minerals. Samples of water draining from eight mines within the District show concentrations of base metals related to this zonation; iron, manganese, zinc, copper, cadmium, and lead occur in the highest concentrations in drainages from the central zone, and decrease outward. These changes are associated with the weathering of pyrite, which is most abundant in the center.

The toxicity of any substance varies with the species under consideration, as well as the organic matter, oxygen, temperature, and combination of pollutants present in the river.[39] Acute toxicity kills organisms outright, whereas chronic toxicity is caused by long-continued exposure to sublethal levels of a toxin. The most toxic heavy metals are mercury, copper, cadmium, and zinc, all of which were introduced to the Front Range rivers as by-products of mining. All creatures absorb heavy metals from their food or directly from the water. Some organisms can excrete the toxins and are generally more tolerant of polluted environments, but other organisms absorb them permanently. Among the organisms that most efficiently concentrate metals are mollusks such as clams and snails, followed by many types of aquatic plants. Mollusks tend to store metals in their tissue and digestive glands, crustaceans such as crayfish in their exoskeleton and hepatopancreas, fish in liver and muscle, and mammals in bone, kidneys, and liver.[40] The metals can inhibit the functioning of enzymes, the proteins produced by cells which facilitate cellular functions, and the metals may damage the gill surfaces of fish and interfere with respiration.[41]

Several studies have examined heavy metal pollution by manganese, iron, copper, zinc, lead, molybdenum, and cadmium associated with historic mining along the Arkansas River, immediately south of the South Platte River basin. A U.S. Environmental Protection Agency Superfund site has been designated at California Gulch, a mining area within the Arkansas River basin. Superfund sites were designated following the

Comprehensive Environmental Response, Compensation, and Liability Act (CERCLA) enacted by the U.S. Congress in December 1980. This act created a tax on the chemical and petroleum industries and provided federal authority to respond to releases of hazardous substances via short-term removals or long-term remedial response actions. In the Colorado mining districts that have been studied in association with CERCLA, such as the Arkansas River, concentrations of metals are highly dependent on flow and tend to increase during high spring runoff. This suggests that these metals are abundant in the fine sediments that are carried as suspended load during high flows. The high spring runoff thus moves metal-laden sediments downstream.[42] Much of the historic mining activity within the Arkansas River basin occurred along headwater streams. Today, the diversity and total abundance of aquatic macroinvertebrates are lower downstream from the junction of contaminated tributary streams, and higher below the junction of clean tributaries. The bottom-dwelling invertebrate communities below the contaminated tributaries are shifted toward more metal-tolerant species, while the fish population is reduced or absent. The invertebrates at polluted sites also have higher concentrations of zinc, cadmium, and copper than those upstream, as do the gut and gill tissue of trout. Gills and gut are the primary route for uptake of heavy metals by fish, and the brown trout at the study sites rarely reached the age of four years or greater, while their condition in all age classes was judged generally poor. Brown trout along the length of the channel downstream from polluted tributaries showed bioaccumulation of copper and zinc in their livers, and chronically high concentrations of these metals are present in the water.[43]

Several of the Front Range rivers share with the Arkansas River the dubious distinction of Superfund status. Superfund sites have been designated throughout the Clear Creek basin, including the Central City mining district, the Argo Tunnel, and Gregory Gulch. At these sites, soils, groundwater and surface waters contain heavy metals including arsenic, cadmium, chromium, copper, and lead. These sites were listed in 1983, and responses thus far have included the construction of retaining walls to contain tailings, the testing of household wells and provision of bottled water, and removal of mercury contamination from an abandoned mine assay laboratory.[44] A particularly innovative response has been the construction of a wetland that is used as a passive treatment for

The first stage in constructing a wetland for passive treatment of acid mine drainage near Idaho Springs, along Clear Creek.

acid mine drainage. Metals carried by water passing slowly through the wetland are attenuated by being adsorbed onto fine-grained sediments that remain in the wetland.[45]

In 1982, the Clear Creek Superfund Site was ranked 174 on a national priority list of 400 sites because of extensive metals contamination of soils, surface, and groundwater as a result of past mining activities.[46] This Superfund site includes Central City, which is presently undergoing rapid development as a result of approval of limited stakes gambling in 1991. The extensive areas covered by poorly consolidated mill tailings and waste rock increase the potential for sinking of the ground under new buildings, and nearly every construction project has had to excavate and properly dispose of soils contaminated with lead, zinc, arsenic, and mercury. As might be expected, however, such issues are not the focus of tourism promotion. Advertisements for Central City highlight the experience of gambling in a scenic mountain town with a colorful history. Little public recognition is given to the negative effects of that history.

Another site in the region, the 4-mile-long (7 km) Argo Tunnel, built between 1893 and 1904, intersects 27 mines.[47] Although the tunnel is no longer in use, it continues to drain these mines, producing an aver-

Table 3.2. Average metal concentrations draining from the Argo
Tunnel during studies in 1973 and 1986

Metal	Average metal concentration (mg/l)	U.S. Public Health Service standard (mg/l)	Average daily contribution (kg)
Iron	340	0.3	910
Manganese	160	0.05	450
Zinc	75	5.0	227
Sulfate	2,700	250.0	8,182

Source: Data from W.H. Ficklin and K.S. Smith, 1994, "Influence of mine drainage on Clear Creek, Colorado." In K.C. Stewart and R.C. Severson, eds., *Guidebook on the geology, history, and surface-water contamination and remediation in the area from Denver to Idaho Springs, Colorado.* U.S. Geological Survey Circular 1097, pp. 43–48.

age summer discharge of 1 cubic foot per second ($0.03 \text{ m}^3/\text{s}$) of water so acidic that has an average pH of 2.8. Iron and manganese are the most serious contaminants to Clear Creek based on U.S. Public Health Service (USPHS) drinking water standards, but copper and zinc also harm aquatic life downstream. Average metal concentrations draining from the tunnel during studies in 1973 and 1986 are listed in the accompanying table. Assuming an average adult weight of 150 pounds, or 68 kilograms, the Argo Tunnel discharge is the weight equivalent of something like 13 times a body weight in iron, 7 times in manganese, 3 times in zinc, and 120 times in sulfate—a lot of metal from a small tunnel. The Argo Tunnel was declared an Environmental Protection Agency Superfund site at the close of the 1990s, and tunnel water is now diverted and treated before entering Clear Creek.

A 1971–72 study of the effects of mine drainage on water quality listed several sites in the Upper South Platte basin that contain levels of metals above those considered safe by the USPHS and the Colorado Department of Health.[48] The sites lie within the South Clear Creek (cadmium, copper, zinc, arsenic), North Clear Creek (cadmium, copper, zinc, arsenic), Boulder Creek (cadmium, zinc), North Fork Poudre River, North Fork South Platte River, and Middle Fork South Platte River basins. The maximum values of zinc, copper, cadmium, manganese, cobalt, nickel, and sulfate, based on 995 stream sampling sites throughout Colorado, occurred at the mouth of Virginia Canyon in the town of Idaho Springs in the Clear Creek basin. Here the levels of these toxic metals exceeded safety standards by several thousandfold.

The Argo Mine and Mill at Idaho Springs, along Clear Creek, in 1995. The Argo Tunnel outlet (center) discharged bright orange water at the time of the photograph. This is now an EPA Superfund site, and water from the tunnel is treated before discharge into the stream channel.

Mining also impacted rivers indirectly by altering the amounts of water and sediment entering the rivers. These alterations were usually primarily caused by destabilization of the valley slopes as a result of lode or placer mining, or as a result of the land-use changes associated with an influx of miners. The valleys of the Front Range are subject to debris flows and landslides under natural conditions,[49] but the frequency and magnitude of these mass movements probably increased during the mining era.

Slope stability is a function of the balance between driving and resisting forces. Resisting forces result from the factors tending to prevent downslope movement of sediment, the friction between individual particles of sediment, and cohesion between particles. Driving forces

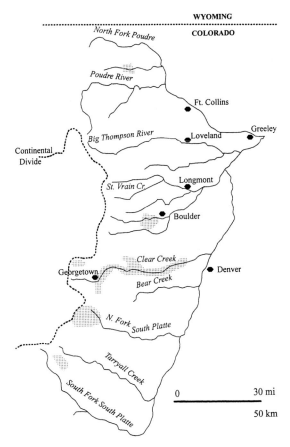

Schematic of stream reaches judged to be affected by metal-mine drainage during a 1971–72 study (shaded). Shaded areas not over a stream course are on channels too small to be shown on this map. (After D.A. Wentz, 1974, *Effects of mine drainage on the quality of streams in Colorado, 1971–72.* Colorado Water Conservation Board, Denver, 117 pp., Plate 3.)

tending to move sediment downslope result from gravity acting on the weight of a mass. When resisting forces are greater than driving forces, a slope is stable. Anything that decreases resisting force or that increases driving force will cause a mass movement of material downslope as the slope attempts to reach a balance among the new controlling factors. A decrease in resisting force may result from an increase in subsurface water that acts as a lubricant, or from fracturing and other weathering processes that reduce the strength of bedrock. An increase in driving force occurs when lateral or underlying support is removed from a slope, or from the addition of mass to the slope.

Activities associated with mining destabilized valley slopes in several ways. Lode mining often removed material from the subsurface, reducing underlying support. Mining wastes that were dumped on the slope as tailing piles below the mine added weight to the lower slopes.

Tailing piles left by a dredge boat operating in the 1920s remain today along the banks of North Clear Creek.

Placer mining along North Clear Creek has removed lateral bank support and enhanced lateral channel movement, producing the undercut banks at right and the linear mound of tailings at left.

An 1898 view of the Gregory Gulch area, showing roads, houses, tailing piles, and treeless slopes. (Photograph courtesy of the Denver Public Library, Western History Department)

Placer mining redistributed material in the valley bottoms, often removing lateral support at the base of slopes. The construction of roads and buildings along the slopes compacted slope surfaces and increased the weight over portions of the slopes. Finally, and probably most importantly, the slopes of the Front Range were largely deforested. Miners needed timber for sluices, flumes, stamp mills, mine timbers, houses and other buildings, and for the fires that drove the steam-operated stamp mills and the smelters. Many of the travelers who noted the muddy stream flow also commented on the denuded slopes.[50] C. M. Clark wrote of the Central City mining district in 1860: "The mountains around these mining districts possess a drear and barren appearance, being covered with but little verdure, and the timber that once covered them being cut off, leaving a multitude of stumps."[51]

Much of the forest destruction in the Front Range occurred through either deliberately or accidentally set fires. En route to the Gregory Dig-

An undated view of Bobtail Hill in the Central City district, showing roads, tailing piles, and treeless slopes. (Photograph courtesy of the Denver Public Library, Western History Department)

gings on 19 June 1859, E. H. N. Patterson wrote: "As we progress, dense forests lie before us; among those forests are raging fierce fires, the smoke of which rises and entirely obscures the sun. All through these mountains we see thousands of acres of dead pines—sometimes whole mountain slopes, with but a green bush to break the monotony of the scene."[52]

Other travelers echoed his remarks.[53] In the days before the U.S. Forest Service was organized to suppress forest fires, on 17 June 1859 a newspaper in the Clear Creek area remarked, "The forests are still burning and will in all probability continue to burn for some time."[54]

Forest fires were sometimes deliberately set to expose bedrock outcrops for prospectors, or to evade timber-use laws. Coal and iron began to be mined in the Front Range in the 1870s, but the coal could not meet the demands of railroads and smelters. Charcoal became the primary smelting fuel between 1875 and 1900, leading to further deforesta-

View of the Caribou mines in the Central City district between 1884 and 1910. (Photograph courtesy of the Carnegie Branch Library for Local History, Boulder Historical Society Collection)

The Boodle Mine along upper Gregory Gulch, 1995. The slopes are being reforested, but large tailing piles and retention ponds (center) remain.

tion. A concerned Congress passed the Free Timber Act of 1878, which prohibited the cutting of live trees on the public domain for commercial purposes. One way to get around these restrictions was to set forest fires that created standing charcoal and dead trees that could then legally be harvested. Such practices persisted until the creation of the first national forest reserves in 1897. When John Tice traveled through the Front Range mining districts in 1871, for example, he noted that in Boulder County alone there were 51 indictments pending for starting such fires.[55]

Several studies have been conducted to estimate changes in forest-fire frequency during the last two centuries.[56] Prior to 1859, fire recurrence intervals averaged 12 years at the base of the mountains, 40–66 years in the ponderosa pine/lodgepole pine/Douglas fir forests of the intermediate elevations, and 300–400 years in the high-elevation spruce and fir forests. These were primarily surface fires at low elevations and the higher-intensity crown fires at the upper elevations. These patterns were altered between 1860 and the advent of fire suppression in 1920. Kirk Rowdabaugh examined fire-scarred trees in Rocky Mountain National Park and adjacent national forest lands. He found a marked increase in fire incidence during settlement, although the fire recurrence interval fell steeply, to an average of 84 years, with the establishment of the national park and active fire suppression. This recurrence interval is 5 times longer than the norm for this type of ponderosa pine–mixed conifer ecosystem, and 5.5 times the recurrence interval in the adjacent national forest lands. A subsequent study of fires in the ponderosa pine ecosystem of upper Poudre Canyon estimated recurrence intervals of 66 years pre-1840, 17 years from 1840 to 1905, and 27 years after 1905. In general, the period from the start of placer mining to the early twentieth century was one of more frequent and widespread forest fires, which would have resulted in destabilization of the slopes, and increases in sediment introduction to the rivers.

Vegetation acts to stabilize slopes in several ways. The above-ground portion of the plants provides a canopy that intercepts precipitation, reducing the erosive impact of raindrops. The surface layer of plant litter, or forest duff, provides a permeable cover on the slope so that precipitation or melting snow filters into the ground more slowly, rather than concentrating on the surface and rushing downslope in erosive sheets or rills. The plant litter also supports soil microbes, which in turn support

Four Mile Canyon, a tributary of Boulder Creek, after a flood in 1894. The railroad tracks crossing the canyon mouth have been destroyed, and a wedge of sediment has been deposited. Note the sparse forest cover on the slopes. (Photograph courtesy of the Carnegie Branch Library for Local History, Boulder Historical Society Collection)

worms and other burrowing animals, all of which increase the ability of the soil to absorb water. The roots of the plants act to bind unconsolidated sediments together, increasing the slope sediment's resistance to movement.

When vegetation is removed by cutting or by fire, all of these stabilizing processes are reduced.[57] In addition, a hot fire may vaporize organic molecules and drive them several inches down into the soil column, where they form a water-repellent layer. This layer is relatively impermeable, so that during an intense rainstorm water does not soak into the slopes, but is concentrated in the upper portion of the soil. Once this upper layer becomes saturated, it can fail catastrophically in a debris flow, as described for the Huachuca Mountains of Arizona, and the coastal ranges of California.

Contemporary accounts describe these processes occurring in the mining districts. Aaron Frost wrote of the year 1872, when "many thou-

Looking up the mouth of Four Mile Creek from Boulder Canyon in June 1995. Relative to the scene in the previous photograph, forest cover has largely returned to the slopes, and a highway (center of photograph) follows Boulder Creek.

sands of tons of rocks, pines, and other debris were washed from Silver Gulch on to the town site of Georgetown, completely covering clumps of mountain aspens to a depth of several feet."[58] A few years later another debris flow covered a local wagon road. Frost also remarked on the "vast aggregations of debris found at the mouths of the lateral gulches," noting that during heavy rainstorms pine trees and other debris were washed into these gulches, forming temporary dams which then burst, flooding and sweeping debris to the foot of the gulch. In his reminiscences, James Rogers describes a summer thunderstorm over upper Clear Creek that washed a mass of debris, talus, and mine dumps from Cherokee Gulch into the valley, burying the town of Brownsville.[59]

The immediate effect of removing slope vegetation is to greatly increase the water reaching a stream channel as surface runoff after a given amount of precipitation, and thus to increase annual and peak streamflows in the channel. Vegetation removal reduces the amount of water lost through evapotranspiration from plant surfaces and may increase the rate of snowmelt.[60] These effects have been documented for streams throughout the Canadian and U.S. Rocky Mountains and the coastal ranges. The effect was also recognized by some of the early inhabitants of the Front Range. H. J. Mattis operated a sawmill at the headwaters of Elkhorn Creek in the early 1880s. In a later controversy with the U.S. Forest Service, Mattis claimed that there was a greater flow of water on the small stream after timber was cut on the basin slopes.[61]

The increases in runoff from deforested slopes were sufficient in some areas to cause river erosion. One study compared river characteristics at 36 sites in the Central City District with those at 31 less-disturbed sites nearby.[62] The Central City sites were characterized by arroyos, or steep-walled trenches with a stream at the bottom. With an increase in runoff caused by the deforestation associated with mining, flooding became a problem for many of the mining towns located along the valley bottoms. As flow depth increases, the force exerted by the flowing water on the channel bed also increases. If the force is large enough to remove the sediments forming the channel bed, the river may erode downward, or incise. In the Central City District, old photographs and sketches indicate that this incision was occurring by the early 1860s. Flow force may have increased by a factor of eight in these streams. Some streams incised 23 feet (7 m) below their premining levels, and at the time of the 1979 study, some of these streams remained unstable.

Over a period of months to years, a more substantial effect of removing slope vegetation is an increase in sediment moving off hillslopes as they become unstable, as documented in studies throughout the mountains of the western United States.[63] In the Front Range, sediment coming off the cleared slopes was probably a mixture of everything from clay to boulders. The finer sediments would increase water turbidity and be carried downstream, causing the problems for aquatic organisms described previously.[64] The gravel, cobbles, and small boulders might be carried short distances downstream, causing channel filling and widening,[65] and perhaps changing the stream planform to that of a braided channel. Streambank erosion would be increased by the removal of stabilizing vegetation,[66] further increasing sediment load. A likely effect of sediment accumulation in channels would be the preferential filling of pools and consequent loss of habitat as described earlier. The coarsest sediment would probably not move very far downstream, creating a riffle or rapid at the point of entry or, in some cases, temporarily damming the channel. In any case, the net effect would be to alter both the total sediment load of the river and the distribution of coarse sediment along the river. These channel changes associated with

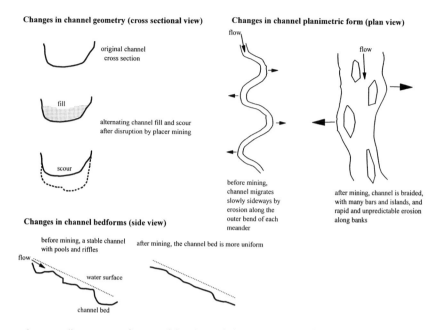

Changes in channel geometry (cross sectional view)

original channel
cross section

fill

alternating channel fill and scour
after disruption by placer mining

scour

Changes in channel planimetric form (plan view)

flow

flow

before mining,
channel migrates
slowly sideways by
erosion along the
outer bend of each
meander

after mining, channel is braided,
with many bars and islands, and
rapid and unpredictable erosion
along banks

Changes in channel bedforms (side view)

before mining, a stable channel
with pools and riffles

after mining, the channel bed is more uniform

flow

water surface

channel bed

Schematic illustrations of some of the channel changes associated with placer mining along the rivers of the Colorado Front Range.

increased sediment from hillslopes have been particularly well studied in the western coastal ranges of the United States and Canada.

Studies of channel morphology in logged and old-growth basins in the western United States have demonstrated that logged basins have rivers with fewer and smaller pools, smaller woody debris, and less fish and invertebrate habitat.[67] Similar characteristics have been documented along mountain rivers in north-central Colorado. Channel changes associated with timber harvest have been documented for mountainous regions of Poland, New Zealand, Australia, Switzerland, China, and Malaysia, among other areas.[68]

Mining was one of the most disruptive activities carried out along the Front Range rivers. By physically disrupting the stability of channel beds and banks, as well as increasing the introduction of sediment from the slopes, the miners caused channel geometry to change, creating episodes in which channels alternately filled with sediment and then downcut. Channels affected by mining also had an increase in braiding and lateral movement and reduced sinuosity. The increased sediment movement disrupted channel bedforms, such as pools and riffles, thus reducing habitat diversity. The increased suspended sediment load from mining stressed or killed aquatic organisms, as did the introduction to the rivers of toxic by-products of mining. These effects of mining were most severe in the southern Front Range rivers Boulder and Clear Creeks, and the Middle Fork of the South Platte River, where the placers and lode ores were richest. At the same time that mining was occurring in the southern rivers, timber harvest and tie drives associated with the expansion of the railroads were impacting the northern rivers.

Timber Harvest and Tie Drives

In 1870 government geologist Rossiter Raymond wrote: "I desire to call attention particularly to one of the worst abuses attendant upon the settlement of the mining regions and other portions of the West. I allude to the wanton destruction of timber." [69] Lumber exploitation closely followed gold and silver mining, because wood was required for sluices, mine timbers, houses, firewood, charcoal, and railroad ties. After a gold strike, lumber mills would spring up overnight. Mining began at the Jefferson Diggings in January 1859, for example, and by April of that year there were 55 miners engaged in sawing timber. By July steam-powered lumber operations had opened near the mining towns.[70] Traveling along the South Platte River from Denver to Tarryall in 1860, Albert Richardson described sawmills, shingle factories, and log houses. Historic photographs of the Front Range from the 1860s to the 1880s, when matched by photographs taken in the 1980s, vividly document the extent of deforestation.[71]

Much of the Front Range lumber went to form ties for the new railroads that brought supplies up to the miners and carried ore down from the mountains. Along with forest fires, this demand probably had the greatest impact on the Front Range forests. Writing of an 1872 trip through the Front Range, William Jones described how "the axe has swept through the mountains and left them a wilderness of stumps." [72] By 1868 the Front Range forests were filled with a small army of laborers —graders, bridge builders, tie and timber choppers, sawmillers, quarrymen, teamsters, and tracklayers.[73] In 1867 the Union Pacific Railroad reached Cheyenne, Wyoming, and in 1870 a branch connected Cheyenne and Denver. The Kansas Pacific also connected Denver to the eastern United States in August 1870, and these larger routes stimulated the development of narrow-gauge lines to the mining centers:[74] in 1870

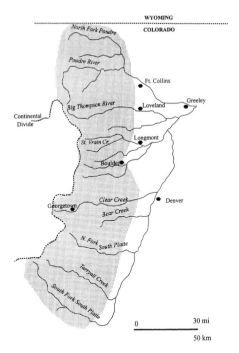

WYOMING

COLORADO

North Fork Poudre

Poudre River

Ft. Collins

Greeley

Big Thompson River

Loveland

Continental
Divide

St. Vrain Cr.

Longmont

Boulder

Clear Creek

Georgetown

Bear Creek

Denver

N. Fork South Platte

Tarryall Creek

South Fork South Platte

0

30 mi

50 km

Shaded area indicates portions of the Colorado Front Range likely to have been affected by timber harvest of some type (for railroad ties, mining supplies, firewood, building materials, or other use) during the nineteenth and early twentieth centuries.

a line reached Golden and was then extended to Black Hawk (1872), Georgetown (1877), and Central City (1878). In 1883 the Union Pacific laid track 14 miles (23 km) up Boulder Canyon, with 66 bridges across the river. Although most of this line was washed out by a flood in 1894, the line was rebuilt and doubled in length by 1898. When another flood in 1919 destroyed the bridges and washed out the track once more, the so-called Switzerland Trail was closed. By 1891 railroad grades existed on both sides of the South Platte Canyon, and two railroad bridges crossed the river.

The Colorado and Southern railroad line reached Leadville from Denver in 1877, running along the North Fork of the South Platte River. By 1879 there were 70 sawmills in the vicinity, and 10 mills on the South Fork taking timber to the town of South Platte. The capacity of these mills ranged from 5 to 60 thousand board-feet (MBM) per day, where a board-foot measures 1 square foot by 1-inch thick. To put this in perspective, the typical yield from ponderosa pines in the Colorado Front Range is 1.2–

An 1884 photograph of the newly completed Georgetown Loop on the Colorado Central Railroad in Clear Creek Canyon. Note the alteration of the valley bottom for the multiple railroad tracks, and the sparse forest cover on the valley slopes. (Photograph by William Henry Jackson, courtesy of the Colorado Historical Society)

A June 1995 view of the same elevated railroad bridge shown above. The perspective is from the other side of the valley, but it is apparent from the more recent view that forest cover has increased, and the valley bottom has been narrowed by the embankment for Interstate 70 (at left).

A sawmill operation in Boulder Canyon during the 1890s. Note the logs in the river at lower right. (Photograph by J.B. Sturtevant, courtesy of the Denver Public Library, Martin R. Parsons Collection, from the Boulder Historical Society)

1.7 MBM per 2.5 acres (1 ha). In other words, even the smallest mills could process 7.5 acres (3 ha) of trees in a day. Lumbermen working for the sawmills or the railroads concentrated on the large trees, but the charcoal burners cut everything, until in some areas they were forced to grub out stumps. Timber was transported by skidding with mules or hauling with horses, and cut wood was transported to the railroad lines with steam tractors or trailers. Timber was also sent to mills in tie drives. The previous section detailed the effects of deforestation on stream channels. Tie drives also altered channel characteristics because the activity occurred within the rivers.

Railroad ties were floated down all of the larger streams from the Poudre River south into the Arkansas River drainage. The main driving streams in the Front Range were the Poudre River, Big Thompson River, St. Vrain Creek, Boulder Creek, and the South Platte River. Many of the Front Range rivers did not carry a sufficient volume of water to float timber except during the annual spring snowmelt floods. The ties cut during fall and winter were often moved to the rivers along flumes, then piled

Horse-drawn logging cart on rails. Logging in the Eldora region of upper Boulder Creek grew with mining in the late 1890s. The loggers shown here worked for the Felch and Jones sawmill, one of the area's largest. Logs were collected on loading docks and transferred to horse-drawn carts (circa late 1890s). (Photograph courtesy of the Carnegie Branch Library for Local History, Boulder Historical Society Collection)

in the bottoms of the channels and extended outward on both sides, with several rows of ties behind each other, stacked to conform to the shape of the creek bed. These stacks often reached a height of 25 ties in the center.[75]

The timing for releasing the ties at high water had to be carefully planned because insufficient water would prevent the ties from reaching their destination, but an unexpected flash flood could scatter broken ties along the river valley. An 1882 flood on the Platte River swept up 25,000 ties held in place by pilings driven at least 11 feet (3.5 m) into the channel bed and washed them downstream past the collection boom, resulting in the loss of 10–15,000 ties.[76] In an attempt to control water flow, splash dams were built along many streams to store water that could then be released at appropriate times to ease the ties through difficult channel reaches.

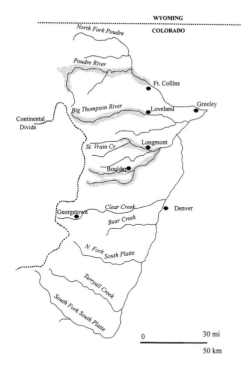

Schematic of the principal stream reaches along which railroad tie drives occurred in the late 1800s. Ties were also floated down portions of the South Platte River near Denver, although the location was not specified by historical sources.

It was common practice to facilitate the downstream movement of ties by altering irregularities along the rivers. Such alterations in the lower gradient or meandering rivers included blocking off sloughs, swamps, low meadows, and banks along wider sections with walls of vertically placed logs. In steeper channel reaches, large boulders, logs, debris, and encroaching streamside vegetation were cut or blasted to facilitate passage.[77] Despite these efforts, ties could still become jammed along the rivers, in some cases creating piles of ties 30 to 40 feet (9–12 m) high with the river running over them in waterfalls. The base of the jam then had to be blasted or broken up with long, spike-tipped poles.

At some point downstream, usually where a railroad crossed the river, a boom was built to catch the ties. Booms consisted of log cribs 6 to 8 feet (2–2.5 m) square filled with rocks, with cables stretched between them like a giant fence. The distance from cutting area to boom varied from a few miles on short streams such as Boulder Creek, to more than

Flumes were used to transport ties to rivers. In this 1916 photograph, ties are piled along the Lava County flume used by Sandford Mills in Wyoming's Washakie National Forest. (Photograph courtesy of the American Heritage Center, University of Wyoming)

Railroad ties stacked along a river of the Medicine Bow National Forest in Wyoming. A splash dam has been built to facilitate tie movement. (Photograph courtesy of the American Heritage Center, University of Wyoming)

Tie hacks (men) wielding pikes to facilitate the movement of ties along a Wyoming stream channel. (Photograph courtesy of the American Heritage Center, University of Wyoming)

50 miles (80 km) from tributaries along the Poudre River to booms at Ft. Collins or Greeley. The tie drives employed anywhere from 30 to 100 men, and an individual drive might last over a month.[78]

The first substantial tie cutting began around Chambers Lake, near the head of Poudre Canyon, in response to construction of the Union Pacific Railroad between Cheyenne and Denver.[79] Cutting began by the spring of 1868, and within two weeks the contractors were delivering 1,200 ties per day. During the winter of 1868–69, more than 200,000 ties were cut and floated down the Poudre, with a similar rate of operations maintained the next year. Greeley, and then Ft. Collins, became the most important tie outlet in northern Colorado, and the Poudre drainage produced most of Colorado's ties from 1874 to 1885, until the bulk of

Members of forestry clubs inspecting ties decked on banking grounds at the Nelson Tie Camp, Colorado National Forest (west of the upper Poudre River basin), June 29, 1926. (Photograph courtesy of the Ft. Collins Public Library)

the trees in the watershed that were at least 8 inches (20 cm) in diameter had been cut.[80] By 1915 tie drives down the Poudre ended, although logs continued to be cut in the region of Chambers Lake and sent west or north. Between 1864 and 1868, the Union Pacific Railroad obtained 38% of the land in the North Fork of the Poudre drainage. Then, from 1870 to 1880, millions of ties were cut, mostly from the lower foothills. The first sawmill in the Poudre basin was a portable one built near the canyon mouth in 1862. By the 1880s there were several portable sawmills in the mountains. The sawmills produced a variety of waste products; there is at least one record of sawmill operators being cited for dumping sawdust into Clear Creek.[81]

Tie drives along the Big Thompson River began in 1874, but this basin had fewer tie drives than the Poudre. An exception was the year 1875, when two Loveland pioneers secured a contract for thousands of ties that they took from the Big Thompson drainage. At least one sign of that single year's work was still evident 82 years later, because on one place on the mountainside above the Forks, "so many ties were slid down the mountain that a deep groove was made and it is still visible, and has always been called the 'Tie Slide'."[82] The Big Thompson drainage did have extensive cutting, however.[83] A sawmill operated near the mouth of the Big Thompson in 1867, and other mills moved up the canyon during the next three decades. Along with cattle raising, the main industry in the area during the 1870s–90s was logging for posts, poles, and other forms of lumber.

The tie drives moved south to St. Vrain Creek in the late 1870s, and the Denver, Western and Pacific Railroad laid tracks up St. Vrain Canyon to facilitate delivery. Seven sawmills were operating in the canyon at one point.[84] Tie cutting and drives were carried out from about 1870 until after 1900 on Boulder and South Boulder Creeks, with most of the cutting done in the 1870s and 1880s, when hundreds of thousands of ties went annually to the Union Pacific Railroad and to smaller lines.

The drainages south of Boulder Creek differed in that river drives were not the primary means of transportation. Between 1869 and 1890 a good many ties were cut in the region, but they were brought out by teams or, later, by railroad. In 1870, for example, the *Rocky Mountain News* contained an advertisement for 500 men with teams to haul 300,000 ties from the forests along the Continental Divide. The railroad also had about 650 ox or mule teams hauling ties. On the whole, more ties were probably cut from the 1880s to the turn of the century than during the preceding 20 years, but the business was no longer as newsworthy, and records became more sparse.[85]

As described earlier, the Front Range rivers were modified to render them more efficient conveyors of the railroad ties. Valley bottoms were diked off from the river channel, obstructions were removed, and the passage of the ties probably had an effect similar to that of running a giant bottle brush down the rivers. No one analyzed these changes at the time, but subsequent research on mountain streams in the Medicine Bow National Forest of southern Wyoming has compared analogous streams that did and did not have tie drives.[86] The streams that had drives tended to be 1–3.6 times wider with minimal bank development, riparian or bank cover for fish, and habitat diversity. They had more riffles, pools that were less well-developed, smaller amounts of large organic debris, and fewer debris-related habitats for aquatic organ-

Ties and saw logs cut from a forest sale area in the Cheyenne National Forest of Wyoming. (Photograph courtesy of the American Heritage Center, University of Wyoming)

isms. The streamside vegetation of these waters was dominated by uniformly aged lodgepole pine (*Pinus contorta*) dating to the time of the last drive, in contrast to the uneven-aged, old-growth vegetation of the streams without drives.

Numerous studies in the Rocky Mountains and coastal ranges of North America have documented the important functions of naturally occurring, large woody debris in mountain rivers.[87] This debris stores wedges of sediment and organic material upstream and thus contributes to substrate diversity and habitat complexity at various scales. The debris creates pools by either causing a step in the channel profile and associated plunging flow, or directing the current toward a portion of the channel bed or bank. These pools form backwaters that provide critical summer and winter habitat and serve as refuges and rearing areas for fish. The debris also provides habitat and food for aquatic invertebrates on which fish feed. A study along the lower South Platte River found higher abundances and diversity of macroinvertebrates on stable wood and boulder substrates than on sand.[88]

In many Front Range rivers, large woody debris was removed dur-

Large woody debris creates pools and protected backwaters that increase habitat diversity along mountain rivers.

ing tie drives. And of course deforestation greatly reduces the standing trees that may eventually fall into a river and form large woody debris. A study comparing 11 undisturbed streams draining unlogged, subalpine old-growth forests in north-central Colorado with four streams draining basins which had been logged prior to 1900 found that the streams in logged areas had significantly less and smaller woody debris, and hence fewer and smaller pools, than did the undisturbed streams.[89] This study indicates the long period of time that may be necessary for a river ecosystem to recover fully after a human disturbance.

Recent removal of large woody debris from mountain rivers elsewhere in western North America supports the inference that the Front Range rivers were dramatically affected by the loss of this debris.[90] The removal of this material decreases pool volume and number, channel stability, the variability of size in sediment forming the channel bed, overhead cover for fish, and travel times of floods, and it alters fish population abundance and species composition. These effects have been observed along rivers with drainage areas ranging from less than 1 square mile (1 km²) to more than 1,500 square miles (4,300 km²). In other words, regardless of river size, the general effect of removing large woody debris is to reduce the natural diversity of river form and function, creating a relatively uniform channel that may efficiently convey water and sediment but that is less able to support aquatic organisms.

In general, the effect of the tie drives on the Front Range rivers was similar to that of placer mining: disruption of bed and bank stability, with consequent increase of sediment movement, increase of lateral and vertical movement of the river channel, and loss of channel diversity and streamside habitat. It is difficult to estimate the relative impact of these two land uses, but the Poudre and Big Thompson Rivers seem to have been more impacted by tie drives, whereas Boulder and Clear Creeks and the South Platte River were more affected by placer mining.

As the Front Range and western Great Plains became more heavily settled, the tie drives created problems for other land users. In 1873 the farmers of Boulder County objected to the destruction of irrigation dams by tie drives,[91] and the loggers were required to post bonds to cover potential damages in Larimer County, which contains the Poudre and Big Thompson Rivers.[92] These conflicts signaled the rise of the next major human activity to affect the Front Range rivers: the construction of reservoirs and flow diversions for agriculture.

A Plumbing System for Crops

Many of the earliest interruptions of flow in the Front Range rivers, whether through diversion of water from the river or storage of water in a reservoir along the river, occurred in connection with mining in the 1860s and 1870s. James Meline wrote of an 1866 journey from Idaho Springs to Central City:

> [we] . . . went up, up, along what had been the bed of a mountain torrent, until we looked almost perpendicularly down on Central [City]. This bed, too, had been torn up, if possible, worse than the lower one. At the top we found water-wheels, the motives for various crushers, and gutters and small aqueducts, led hither and thither, up and down the mountain, at the behest of the 'Consolidated Ditch Company'—I believe that's the name—an enterprising corporation that, by some legislative arrangement, have managed to monopolize all the water in these mountains, and sell the privilege of its use at fabulous prices.[93]

Simultaneously, the rapidly growing towns at the base of the mountains developed diversion and reservoir systems for both agricultural and municipal purposes.[94] Among the earliest were ditches built in 1859 from Clear Creek to Golden, and along Bear and Boulder Creeks. These were

Table 3.3. Selected water diversion and reservoir systems in the Colorado Front Range (Sources Gerlek, Tyler, Wroten)

System	Construction dates	Location and purpose
South Lone Pine Ditch	pre-1879	
Highline Canal Intake Dam	1879	Eighty-four miles (140 km) long; South Platte Canyon through Denver; irrigation
Chambers Lake Dam	1891	Poudre Canyon; storage
Skyline Ditch	1891–95	N. Platte and Laramie Rivers to Chambers Lake
Big Dam	1882, 1895	Lower Big Thompson; storage
Grand River Ditch	1890–1929	Diverted water from the Colorado River to the Poudre River
Argo Tunnel	1893–1910	Four-mile (7 km) tunnel to carry water from Central City to Idaho Springs (S. Fork Clear Creek)
Barker Dam	1909–1910	Boulder Canyon; hydroelectric
Buttonrock Dam	1966–1969	St. Vrain; storage for Longmont

Big Thompson Project
  Carter Lake Reservoir
  Horsetooth Reservoir
  Estes Park Aqueduct
  St. Vrain Supply Canal
  Boulder Supply Canal
  St. Platte Supply Canal
  Poudre Supply Canal
  N. Poudre Supply Canal
  Horsetooth Feeder Canal
	1938–1956	Northern half of S. Platte River basin; irrigation and storage

Laramie-Poudre Tunnel	1911–1914	Laramie River to Poudre River; irrigation and storage
Laramie River Feeder Ditch	1891	As above
Cameron Pass Ditch	1913	N. Platte to Joe Wright Creek; irrigation

As of 1977, there were more than 93 reservoirs in the Poudre basin; although only approximately 12 of these were in the mountains.

followed in 1860 by ditches on the Poudre River, St. Vrain Creek, and the South Platte River below Denver; in 1861 by diversions from the Big Thompson River; and by Denver's City Ditch, dug in 1865. Most of these ditches were privately owned by single farmers and were large enough to irrigate only a few acres of land. The Union Colony (later Greeley) initiated the cooperative construction of irrigation canals in the Front Range in the 1870s with the Greeley Canal Company No. 2 Canal. Similar irrigation communities were quickly established at Ft. Collins, Loveland, and Longmont.

Irrigation ditches rapidly spread through the mountains as well. For example, in the Red Feather Lakes District of the North Fork Poudre River, the South Lone Pine Ditch in 1879 inaugurated several ditches that by 1900 had converted numerous swampy depressions into lakes.[95] The Left Hand Ditch diverted waters from St. Vrain Creek for irrigation in another basin.

The ditches grew rapidly in length. The Highline Canal Intake Dam, 2.5 miles (4 km) upstream from the mouth of South Platte Canyon, was begun in 1879. It served as the source for an 80-mile-long (140 km) canal that ran through Denver and provided irrigation water.[96] In 1883 William Pabor described the Poudre River valley immediately east of the Front Range: "From LaPorte to its junction with the South Platte, thirty miles below, the Poudre Valley is one vast network of irrigating canals, mainly taken out upon the north side of the stream. It was in this valley, in 1871, that the completion of an irrigating canal of the Greeley Colony, and its successful working, gave the first impetus to farming in Colorado. . . . Nearly a score of canals, varying in length from ten to thirty miles, utilize the water from this stream [Poudre], and others are in course of construction."[97]

Pabor described the primary canals serving the Greeley area, which was called the Garden Town of Colorado, or Forest City: Greeley Canal (1871), Lake Canal (1872), Mercer Ditch (1862), Box Elder Canal (1863), Cache-la-Poudre Canal (1866), Canal Number Two (1872), Pleasant Valley Canal (1880), Larimer County Canal (1872), North Poudre Canal (1880), and Larimer and Weld County Canal, which was the largest canal in operation in Colorado in 1883. In aggregate, these canals extended well over 100 miles (170 km) in length. The canals varied between 11.5 and 15 feet (3.5–9 m) in width, and 2 and 5 feet (0.6–1.4 m) in depth. They in turn supplied thousands of smaller ditches and laterals.

A diversion structure along North Clear Creek. The headgate is at center right. The channel is also constrained by a road embankment at left.

Elevated wood diversion canal along North Clear Creek, 1995.

View of the Highline Canal in lower South Platte Canyon.
(Photograph courtesy of the Colorado Historical Society)

At the same time, dams were being built to ensure a water supply for the canals.[98] The earliest was the 1869 reservoir on Ralston Creek, a tributary of Clear Creek. Later examples include the Big Dam, starting point of the Home Supply Ditch, which was built in Big Thompson Canyon in 1882. Chambers Lake (1891), Halligan Reservoir, Eaton Reservoir, and Seaman Reservoir were built along the Poudre River and its tributaries. Together, these reservoirs hold approximately 24,000 acre-feet (34,179,000 m^3), or 12% of the Poudre River's average annual runoff. The various canals and reservoirs converted the natural drainage system to a vast plumbing scheme. Instead of corridors of streamside vegetation following topography and separated by semiarid steppe or montane forest, the region became a network of taps and tubes to serve

A 1995 view of the diversion intake structure and canal beside a dam at the mouth of Big Thompson Canyon.

A 1917 view of the dam just below the mouth of Big Thompson Canyon. (Photograph courtesy of the Colorado Historical Society)

The spillway at Chambers Lake in upper Poudre Canyon (circa 1904). Note the extensive deforestation in the background. (Photograph courtesy of the Ft. Collins Public Library)

human activities such as mining or agriculture. Water could thus be transferred from wetter to drier areas, regulated according to season or hour, and even transferred across drainage divides thousands of feet above sea level.

Irrigation was the impetus for the early water projects not tied to mining. As noted in the first chapter, the Front Range is semiarid, receiving approximately 15 inches (40 cm) of mean annual precipitation at the mountain front. Superimposed on this mean value are dramatic annual to decadal fluctuations that have brought from 7 to 28 inches (18–70 cm) of annual precipitation; consequently, rainfall for crops is unpredictable and unreliable. But the Front Range rivers always flow, even in drought years, so the farmers and ranchers who followed the miners began to use the more dependable streamflow rather than relying on rainfall. In 1880 Frank Fossett wrote that "all through the northern part of the mountain sections are fertile valleys and grazing lands. . . . There

The Laramie River Ditch, which transfers water from the upper Laramie River across Green Ridge to the upper Poudre River. (Photograph courtesy of the Ft. Collins Public Library)

are countless farms or ranches all through the mountains, some of them of considerable extent and of great fertility. In the little mining county of Gilpin, the annual farm and dairy products must exceed seventy-five or eighty thousand dollars. . . . There are similar farms all through the hill country of Boulder, Jefferson, and Park Counties."[99]

Many of the farmers in both the mountain valleys and along the base of the foothills dug small irrigation ditches from the nearest river. As settlement density increased at the edge of the foothills, the availability of water became a limiting factor. The establishment of the Ft. Collins Agricultural Colony upstream from Greeley in 1872 led to the construc-

The Grand Ditch was built along the eastern flank of the Never Summer Range in Rocky Mountain National Park to divert water eastward from the Colorado River. Here the ditch forms a scar readily visible across the Colorado River valley.

tion of two irrigation canals that had the combined capacity to divert the entire flow of the Poudre River during low-flow years. Predictably, the dry summer of 1874 caused the Greeley irrigators a great deal of anxiety and ultimately led to the incorporation of the doctrine of "prior appropriation" in the 1876 Colorado Constitution.[100]

Prior appropriation is a legal concept stipulating that the first person to appropriate water and apply it to a "beneficial" use, usually agricultural or industrial, has the first right to use that amount of water from that source.[101] Each subsequent user may claim a portion of the water remaining only after all senior water rights are fulfilled. What prior appropriation does not stipulate is proximity to the water source. If a senior user has the capital to divert and transport water over long distances, then such a use is perfectly acceptable. In 1882 the Left Hand Ditch furnished the Colorado Supreme Court with the first test case of the prior appropriation doctrine when water users within the St. Vrain drainage argued that those within the natural drainage basin had a better right to the use of its waters than did earlier users outside the basin. The Supreme Court denied this assertion.

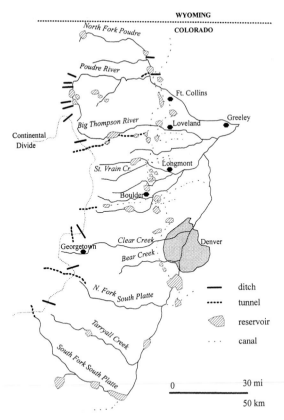

Front Range schematic indicating the principal water reservoirs, tunnels, and ditches across the Continental Divide, and irrigation canals. Because of the scale of this map, not all structures are shown. (After D. Tyler, 1992, *The last water hole in the West: The Colorado-Big Thompson project and the Northern Colorado Water Conservancy District*. University Press of Colorado, Niwot, 613 pp.; J. Meiman and G. Leavesley, 1974, *Little South Poudre watershed climate and hydrology 1961–1971: Synopsis*. Colorado State University, Ft. Collins, 22 pp., Figure 1; and S. Gerlek, 1977, *Water supplies of the Platte River basin*. M.S. thesis, Colorado State University, Ft. Collins, 798 pp., Figure D.)

The farmers settling the eastern slope of the Rocky Mountains quickly realized that the waters of the wetter, less-densely settled western slope were available for appropriation if they could be diverted to the eastern slope. This required more labor and capital than previous efforts, however, and irrigation companies began to form. Early examples in the northern Front Range include the Larimer County Ditch Company (1890), the Water Supply and Storage Company, and the North Poudre Irrigation Company (1881). In 1882 the Cameron Pass Ditch was the first diversion across the Continental Divide in Colorado, to be followed by the Grand Ditch (1892) and the Skyline Ditch (1893). The former diverted western slope streams 15 miles (25 km) north and across the Continental Divide at Poudre Pass and into the headwaters of the Poudre River and Long Draw Reservoir. In 1922 the Poudre River had an annual flow of 340,000 acre-feet (444,060,000 m^3) and transbasin additions of 35,000 acre-feet (43,172,500 m³). In other words, nearly 10% of the annual flow of the river came from outside the natural drainage area.

Increased availability of water stimulated agricultural growth, which in turn required additional water. Feedlots along the plains portions of the major tributaries of the South Platte River increased the demand for corn, alfalfa, and hay, all grown with irrigation. The expanding sugar beet industry of northern Colorado required water both for growing beets and for refining the sugar; by 1933 the Great Western Sugar Company had 17 processing plants in northeastern Colorado. In 1899, 446,300 irrigated acres (180,706 ha) were being farmed in the five counties (Larimer, Weld, Morgan, Logan, and Sedgwick) that form the northern portion of the South Platte basin. This represented an increase of 85% over the previous decade. By 1909 irrigated acreage in the region had increased an additional 68% to 749,150 acres (303,300 ha). Simultaneously, the region's population grew 57.8% percent in the first decade of the twentieth century. During the 20 years between 1890 and 1910, ditch and reservoir construction proceeded at an equally lively pace, but by 1910 the total dependable water supply was already in use, and junior appropriators sometimes experienced water shortages. (Population growth and increase in irrigated acres decreased substantially, and a regional drought during the 1920s and early 1930s further stressed existing systems.) By 1933, 390,000 acres (157,895 ha) of land connected to the ditch system remained fallow due to lack of water, although 47,000 acre-feet (57,974,500 m³), or approximately 4% of the upper South Platte River's runoff, were being diverted from the Colorado and Laramie Rivers.[102] The construction of new reservoirs would make the stored water too expensive for irrigation, so the water planners turned again to the Colorado River and the western slope.

The Northern Colorado Water Conservancy District, formed in 1937, was instrumental in facilitating the Colorado-Big Thompson Project. Constructed between 1938 and 1956, the project was created to supplement the water supply for 615,000 acres (248,988 ha) of irrigated land in northeastern Colorado by diverting water from the Colorado River basin to the South Platte River basin.[103] This project represents the culmination of water movement within the South Platte basin. As of 1992, the eastern slope users had constructed 37 transmountain diversion projects that collectively removed more than 650,000 acre-feet (801,775,000

Table 3.4. Distribution of native and imported surface water in the upper South Platte River basin. (After Gerlek, 1977, Tables 2-2, 2-3, and 2-6.) Allocated water refers to that decreed in water rights; these amounts are to the junction with the South Platte River. Amount of water actually diverted includes much water that is re-used more than once. I is the ratio of allocated water:actual flow, II is the ratio of allocated water:diverted water, III is the ratio of actual flow:diverted water.

Sub-basin	Allocated flow (ac-ft/yr)	Storage (ac-ft)	No. of reservoirs	Actual flow (ac-ft/yr)	Amount of water diverted by use in sector in 1970 (ac-ft/yr)			I	II	III
					Agricultural	Municipal	Industrial			
S. Platte mtns., foothills	4,959,400	393,000	44	213,552	646,235	218,754	63,057	23.2	5.3	0.2
Bear Creek	844,908	31,000	16	44,927	1,054	2,100	0	18.8	268	14.2
Clear Creek	3,583,800	113,000	47	173,994	4,644	2,200	15,191	20.6	163	7.9
Boulder Creek	3,982,000	49,300	31	122,832	108,493	17,894	51,876	32.4	22.3	0.7
St. Vrain Creek	2,200,960	42,200	55	117,600	125,610	11,241	2,613	18.7	16	0.8
Big Thompson River	1,976,520	101,000	14	114,600	229,054	9,792	1,007,715	17.2	1.6	0.1
Poudre River	4,662,560	200,000	76	232,833	493,526	29,048	8,854	19.8	9	0.4
Total	22,210,148	929,500	283	1,167,942	1,608,616 (53%)	291,029 (10%)	1,149,306 (37%)			

m³) of water from the Gunnison, San Juan, and Colorado River basins. The Colorado-Big Thompson Project is the largest of these diverters. Out-of-basin imports of more than 370,000 acre-feet/year (456,395,000 m³) account for an average 32% of the surface runoff in the upper South Platte basin; 94% of the imported water comes from the Colorado River basin. (Deliveries have accounted for an average of 34% of all water used in Larimer, Weld, and Boulder counties during drought years between 1958 and 1990.) Agricultural uses have been gradually supplanted by municipal demand, but they remain high. In 1987, 85% of Colorado's out-of-stream water use still went to irrigation.

The changes in flow regime as a result of water diversion to and from various Front Range rivers are illustrated in the table of water distribution. A comparison of actual flow to surface-water diversions for the channels in this table show that the Big Thompson River has the lowest ratio of natural to diverted flow, indicating a highly altered flow regime. Clear and Bear Creeks have much higher ratios of actual flow relative to water diverted. Boulder Creek, with the highest ratio of allocated to actual flow, has the potential to be the most over-appropriated in the future, but this ratio greatly exceeds 1 for all of the Front Range rivers. At present, Bear Creek has the most allocated but so far undiverted water, and the Big Thompson River has the least.

The flow regime of a river under natural conditions should correspond from year to year to the precipitation falling on the river basin, unless the river has a strong input of groundwater flow. Rivers with many diversions show a disconnection not only between the amount of precipitation and total annual flow, but also between precipitation and minimum flows, which are reduced during the irrigation seasons by diversions. Maximum daily flows mostly track snowmelt and constitute a large proportion of the total annual flow, whereas minimum daily flows represent lows resulting from diversions. The Poudre River illustrates this situation, as it has many diversions and no statistical correlation between precipitation and streamflow levels. In contrast to the Poudre River, Clear Creek has few diversions and strong correlations between the year-to-year variability of precipitation and streamflow. In essence, water use drives the flow regime of a river with substantial diversions, primarily by reducing minimum flows.

The numerous diversions and dams have produced readily recognizable changes in the Front Range rivers. Reminiscing in 1937 about pio-

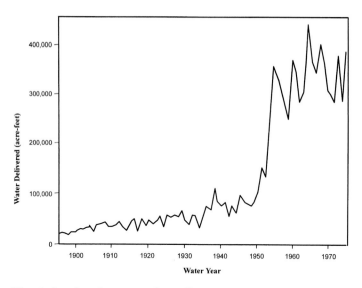

Historical total yearly imports of out-of-basin water to the South Platte River basin, 1895–1974. (After S. Gerlek, 1977, *Water supplies of the Platte River basin.* M.S. thesis, Colorado State University, Ft. Collins, 798 pp., Figure 2–7.)

A pipe that forms part of the Colorado-Big Thompson Project crosses the mouth of Big Thompson Canyon.

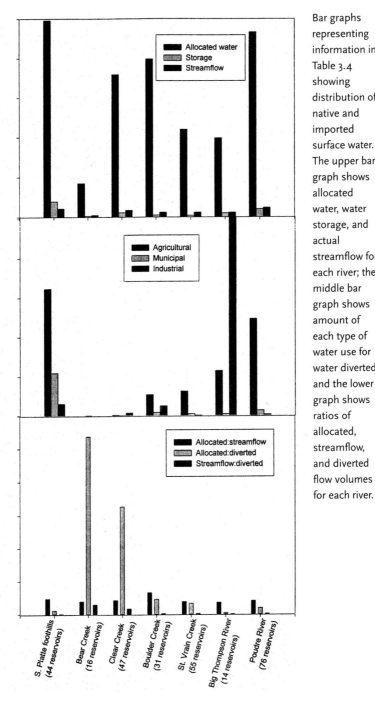

Bar graphs representing information in Table 3.4 showing distribution of native and imported surface water. The upper bar graph shows allocated water, water storage, and actual streamflow for each river; the middle bar graph shows amount of each type of water use for water diverted; and the lower graph shows ratios of allocated, streamflow, and diverted flow volumes for each river.

The snowy peaks of the Colorado Front Range, seen across the farm fields of the western Great Plains.

neering along St. Vrain Creek, Alonzo Allen wrote: "Possibly it would be hard for many people to realize it now, but it seems that the St. Vrain used to be a river in fact as well as in name. As you gaze upon the small trickle of water which ordinarily makes its way down the course south of Longmont, you would scarcely believe that at one time a ferry was kept in readiness for use at such times as the raging St. Vrain should choose to inundate or wash out the bridges. This was, of course, in the days before man had made provision for utilizing just about all the water given to the stream by the mountain snows and rains." [104]

Similarly, in 1959 irrigators along Lone Pine Creek in the North Fork Poudre basin took almost all of the water from the stream during irrigation season, although subsurface seepage and return surface flow from the irrigated fields kept water in the stream at all times. The whole North Fork's entire flow was reserved on water rights in existence at that time. [105]

The effects of water diversions and reservoirs on Front Range rivers result from reduction in flow, augmentation of flow, and storage of water. As explained previously in the discussion of mining impacts, reduction or augmentation of channel flow could affect sediment transport, stream stability, and stream biota. The natural seasonal flow of

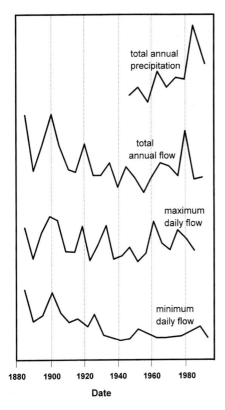

Changes in flow characteristics through time as recorded by the discharge gauge at the mouth of Poudre Canyon (lower three curves). Points represent flow values at selected five-year intervals. Total annual flow and maximum daily flow have not changed appreciably over the period of record. Minimum daily flow, however, drops abruptly after 1925. Stream discharge generally bears little relation to regional precipitation as recorded at the Walden station, which is just outside the Poudre drainage basin, but has the longest continuous record of any station in the region. (A statistical t-test indicated no significant correlation between the discharge measures and precipitation.) The Poudre River presently has nine diversion structures and 32 canal systems, including the Poudre Valley Canal (1901), a half mile (0.8 km) upstream from the discharge gauge. Vertical scale for each curve differs, but represents the magnitude of the quantity with which the curve is labeled.

the Front Range rivers varies with elevation; rivers above approximately 7,600 feet (2,300 m) have a broad snowmelt peak in May-June, whereas those at lower elevations have the same peak, but with short, steep, thunderstorm-driven floods occasionally superimposed in July and August.[106] These flow characteristics are altered in varying ways, depending on whether the river is used as a source or receiver of diverted water, and

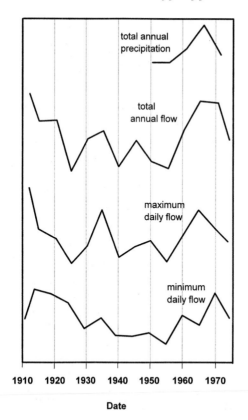

total annual
precipitation

total
annual flow

maximum
daily flow

minimum
daily flow

1910 1920 1930 1940 1950 1960 1970

Date

Changes in flow characteristics through time in Clear Creek (points represent flow values at selected five-year intervals). Clear Creek has relatively little water development upstream from the discharge gauge at Golden; there are four minor import tunnels and the Welch Ditch upstream. Unlike the Poudre River, there are strong correlations between precipitation and annual discharge characteristics (minimum daily flow and total annual flow) in the Clear Creek basin. Discharge data from Clear Creek near Golden; precipitation data from Idaho Springs, in the Clear Creek drainage basin. Vertical scale for each curve differs, but represents the magnitude of the quantity with which the curve is labeled.

on whether the river is dammed. A study of five watersheds in and adjacent to the Front Range revealed that irrigation diversions did not significantly alter peak rates of runoff for watersheds of less than 1,000 square miles (2,800 km²) in which thunderstorms are the primary causes of floods.[107] Flow diversions did become significant for larger watersheds in which annual peak runoff was produced by snowmelt.

Channel changes resulting from alteration of the flow regime vary with channel type.[108] A study of nine rivers with a range of drainage-

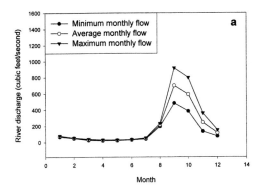

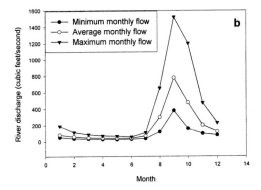

Representative hydrographs of a Front Range channel dominated by snowmelt (a) and by a combination of snowmelt and rainfall (b). Months proceed from October (1) to September (12). Clear Creek at Idaho Springs (a) is at an elevation of 7,600 feet (2,300 m), and the maximum monthly flows during summer are not much larger than the mean monthly flows. Clear Creek at Golden (b) is at an elevation of 6,000 feet (1,830 m) and has large maximum monthly flows in summer caused by thunderstorms, as well as snowmelt-driven flow.

basin areas found that laterally unconstrained, pool-riffle channels from which flow was diverted decreased in width by 30–50% as a result of vegetation encroaching on the channel. Laterally constrained, step-pool channels had no observable change. Flow had been diverted from 20 to 100 years from the rivers in this study. Annual water yield was reduced an average of 40–60% by diversion, although floods that occurred on average once every 5 to 10 years were sometimes released during wetter years.

When river flow ponds behind a dam, sediment in transport settles as flow velocity decreases. As a result, the channel fills with sediment at the dam and immediately upstream, and less sediment is transported downstream. Water released from the dam tends to be cold and clear, and flow volume may fluctuate more rapidly than under natural conditions. This rapidly released water of low sediment concentration may enhance channel erosion immediately below the dam. Following the 1979 completion of a dam on Bear Creek, channel width increased by an average of 119%, flow depth increased by an average of 156%, and channel gradient decreased by 25% relative to pre-dam conditions.[109]

Another effect of dams is decreased water temperature downstream. The elimination of vital temperature cues may retard normal growth of river insects and fish and drastically curtail or prevent insect emergence.[110] A study of rivers below 17 reservoirs in the Colorado Rockies, including four in the Front Range, demonstrated that daily and seasonal fluctuations in water temperature were reduced, and that the seasonal maximum temperature was delayed downstream of reservoirs that release water from the base of the dam. These changes caused a reduction in macroinvertebrate diversity and in the number of taxa.[111] Dams may also release waters low in oxygen, impacting the ability of macroinvertebrates and fish to live in the channel until the water becomes re-oxygenated as it flows downstream. Finally, upstream riffle habitats may be lost when they are covered by water behind the new impoundment. The effects of a dam will depend in part on its structure and function: deep-release dams suppress natural variation in water temperature below the dam; hydroelectric dams have extreme short-term fluctuation in rates of flow release; and storage reservoirs dampen short- and long-term extremes of flow. As a whole, species diversity of aquatic macroinvertebrates is moderately reduced below dams in Colorado; some species such as stoneflies may decline, while others such as dipterans may increase.[112]

Rates of water-level fluctuation particularly affect fish and aquatic invertebrates. These creatures use the *varial zone,* the area of a stream channel that is flooded during spring snowmelt and then dewatered annually over the course of an average water year along the Front Range rivers. Both the annual snowmelt floods and the summer thunderstorm floods recede over periods of days to weeks, influencing the varial zone in a manner for which the biota are adapted. Studies along regulated rivers in the Rocky Mountains, particularly with flow impoundments operated for hydropower, indicate that the varial zone is rapidly and un-

Coarse sediment accumulation above this dam at the mouth of Big Thompson Canyon is visible even at high flow as shown here at center right.

Water released from the base of the dam is cold, turbulent, and depleted of sediment. The dam also forms a barrier to migration of fish. This dam at the mouth of Big Thompson Canyon releases water by spilling from the top of the reservoir.

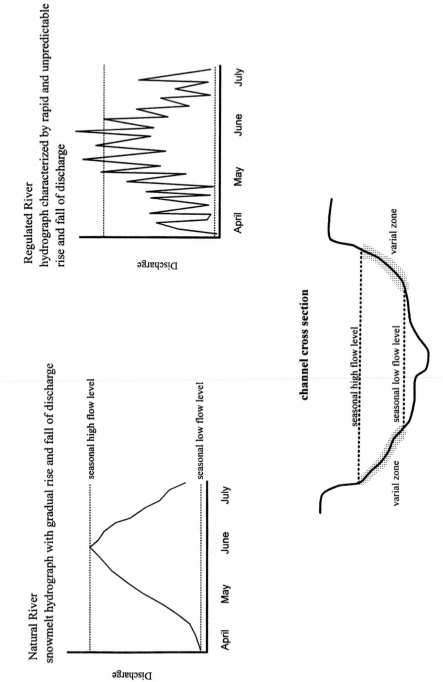

Natural River
snowmelt hydrograph with gradual rise and fall of discharge

Regulated River
hydrograph characterized by rapid and unpredictable rise and fall of discharge

channel cross section

Schematic of the differences in the seasonal hydrograph, and inundation of the varial zone, along a natural and a regulated river.

predictably flooded and dewatered such that the biota have little chance of naturally colonizing newly flooded areas or evacuating dry areas.[113] As a result, insects and fish stranded by a sudden drop in water level may be frozen or dessicated, and aquatic biodiversity may be reduced relative to unregulated river reaches.[114]

The distribution and characteristics of streamside vegetation on the floodplain may also be altered by changes in flow regime, as indicated by studies in Sierra Nevada, the European Alps, and the Rocky Mountains.[115] Studies on a wide variety of channel types demonstrate that various conditions control seedling establishment and tree maintenance: water availability; frequency, duration, and intensity of floods; and fluxes in sediment deposition and erosion.[116] Numerous studies have demonstrated a reduction in both abundance and diversity of streamside forests downstream from dams in both Europe and North America.[117] To be established successfully, cottonwoods, poplars, willows, and other species require bare, moist surfaces protected from disturbance—in other words, the type of surface that may be created by deposition during the waning stages of a flood. When a dam reduces flooding and sediment transport, channel narrowing or other river adjustments will alter the patterns of streamside vegetation establishment.[118]

Dams may also cause an increase in sediment concentration. During the construction of Joe Wright Reservoir on the upper Poudre River in 1978–1979, values of mean monthly suspended solids increased from two to thirty times those upstream, and macroinvertebrate density decreased by up to 90%.[119] When reservoirs begin to fill with sediments, they must be flushed to maintain storage capacity. Flushing of sediments from Dry Creek Reservoir along the North Fork of the Poudre River increased discharge and suspended solids through scouring of the channel bed downstream, and decreased the level of dissolved oxygen. As a result, algal abundance decreased more than 90% and did not recover for 9 months. Invertebrates decimated by the initial release might have recovered in 2–3 months, but further releases reduced recovery rates and shifted the community composition to small, fast-growing species.[120] Another example comes from a sediment release from Halligan Reservoir on the North Fork Poudre River. An estimated 7,000 cubic yards of clay-to-gravel-sized sediment released from the reservoir in late September 1996 partially filled pools and coated riffles for a distance of

6 miles (10 km) below the reservoir.[121] The sediment release eliminated most of the bottom-dwelling macroinvertebrates along the upper portions of the affected reach of river and caused a shift in population structure from scrapers to collector-gatherers at sites along the lower portions of the reach. Recolonization of the affected sites occurred as sediment was gradually flushed downstream, but macroinvertebrate abundance and structure had not recovered to predisturbance levels two years after the release.[122] In general, the ability of an aquatic community to recover depends on the magnitude and frequency of disturbance and on the presence of organisms in unaffected channel reaches that may recolonize the disturbed area.

Water flow may be maintained at an unnaturally high level throughout the growing season in rivers receiving flow diversions, reducing important quiet-water habitat for juvenile fish, as along the Flathead River of Montana, for example.[123] Because most irrigated agriculture occurs along the Plains portion of the South Platte River basin, the Front Range rivers are used as water conduits; that is, their natural flow is more likely to be augmented than depleted. The greatest agricultural demands for water occur during July and August, so the natural snowmelt flow peak is prolonged throughout the summer to ensure water delivery to the Plains. As the snowmelt peak begins to recede along the Poudre River, for example, reservoir storage in the upper basin is used to augment water flow. Later in the summer, the flow is augmented by water diverted from the Colorado River via the Colorado-Big Thompson Project, which is delivered along the lower portion of the mountain segment of the Poudre River.[124] In addition to prolonging the peak flow, artificial flow augmentation also reduces flow variability between years.

Most of the Front Range rivers do not receive surface return flow from irrigated fields within the mountain channel segments because most agricultural irrigation occurs on the plains beyond the mountain front. Where return flow does occur in Front Range rivers, the water reentering the river is generally warmer than water coming from upstream and may carry greater loads of fine sediments. The water may also carry agricultural pollutants, including pesticides, mineral nutrients, salts, radionuclides from fertilizers, wastes from polyethylene tarpaulin and petroleum mulches, the by-products of combustion of fossil fuels, and organic nutrients such as nitrogen from animal manure.[125] Again, this return flow can adversely affect aquatic biota.

In summary, the addition of water diverted from the western slope of the Rocky Mountains to the Front Range rivers has undoubtedly affected the aquatic biota of these rivers. Although few studies have documented these effects in the Front Range, it seems reasonable to draw analogies from other, more intensively studied, mountain rivers in the western United States. Similar considerations apply to dams along the Front Range rivers. There is less reason to believe that the rivers have been physically altered by augmented flow; at the same time, we can speculate that the increased flow, by increasing sediment transport, has enhanced channel recovery from the effects of mining and deforestation. Naturally occurring Front Range river characteristics include relatively low suspended sediment loads and channels formed in boulders and bedrock resistant to transport under increased flow, and both have probably minimized the physical effects of dams. Finally, reductions in flow have undoubtedly altered both stream biota and river form, in the latter case by reducing sediment transport capability. Following the argument suggested previously, this reduction in flow may have diminished river recovery following mining and deforestation.

Roads

Many of the first roads into the Front Range drainage basins were built in association with mining.[126] When C. M. Clark visited the Front Range in 1860, he described a network of four primary roads connecting Central City, Golden, and smaller mining towns. By 1867 the main road connected Georgetown and Silver Plume, and lesser roads snaked throughout the region. Most of these were toll roads because the labor of constructing and maintaining a road was often justified in terms of a business. The earliest roads were often slightly improved tracks from which the trees had been cleared, but they formed the basis for today's paved highways. The early tracks might be improved to corduroy roads by laying logs side by side to reduce the effects of mud and boulders. Along with the roads came the improvements necessary to get them through the narrow mountain valleys: numerous bridges; supporting log structures filled with soil and rock; and tunnels or ledges blasted out of the bedrock. Describing an 1874 trip along Clear Creek on the Colorado Central Railroad, Charles Harrington wrote that "huge boulders, torn from the mountainsides by the engineers who piloted this little road up here,

Table 3.5. Selected Front Range roads

Location	Date
Poudre Canyon	
Road to Chambers Lake	1879–80[1]
Road up much of Canyon	1912–16[2-4]
Narrows blasted	1920
Complete to Walden	1926
Up Boulder and South Boulder Creeks	begun in 1865[5]
Up South St. Vrain Canyon	1894[6]
Two-lane road up Boulder Canyon	1915[7]
Auto stage line from Loveland to Estes Park	1907[8]
Public road up Middle St. Vrain	1910[9]
(partly destroyed by rockslide in 1912)	
Big Thompson	
Toll bridge and stage line	1858[10]
One-lane dirt road with corduroy bridges	1902–04
Improved, widened, & reinforced	1908, 1925-autos
Road paved	1937
Colorado highway 34	1938
South Fork Big Thompson	early 1900s[11]

Sources: 1. Watrous; 2. R.L. Giddings. 1984. *Ipswich to Ft. Collins: The Giddings family, 1635–1985.* Parkview Publishing Company, Ft. Collins, 176 pp.; 3. N.W. Fry; 4. A. Albrandt and K. Stieben; 5. P. Smith; 6. L. Noble; 7. S. Pettem; 8. E.A. Mills; 9. M.D. Dunning, ed. 1975. *Historical reminiscing in the Allenspark mountain area.* Privately printed, 40 pp.; 10. D. McComb. 1980. *Big Thompson: Profile of a natural disaster.* Pruett Press, Boulder, 188 pp.; 11. H.M. Dunning, v. 2

dam up the creek at various points." [127] By May 1871 there were 33 bridges in Boulder Canyon alone. The table listing selected Front Range roads provides an indication of the pace of road building in the Front Range canyons.

Often the original, heavily used roads were abandoned when railroad lines reached an area, only to be replaced later by similar roads. For example, an 1867 road over Loveland Pass fell into disuse when the railroad reached the pass. Yet the High Line Road over the pass, completed during the winter of 1878–79, was used by up to 50 teams a day by early June 1879. From 1911 to 1922 an automobile road was built along a similar route, and in 1949–50 the road was paved. The adjacent 1.7-mile (2.8-km) Eisenhower Memorial Tunnel (constructed 1968–1973)

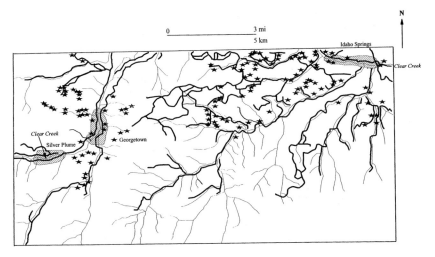

Detail map of the Georgetown-Idaho Springs area, illustrating the density of perennial and ephemeral channels (light lines), roads (heavy lines), and mines (star). The principal areas of each settlement are shaded. (After Spurr and Garey, 1908. U.S. Geological Survey Professional Paper 63, Plate XVII.)

enabled motorists on Interstate Highway 70 to bypass this slow, often treacherous route, and fifteen million vehicles passed through the tunnel within four years of its completion.[128]

Roads were built along primary transportation corridors, to individual mines and ranches, along large irrigation ditches, and into areas of timber harvest. The effects of these roads varied, depending on the type of construction, and fall into four general categories: increased hillslope sediment movement from road surfaces, decreased slope stability and concomitant slope failures, changes in river form associated with confinement or stabilization by bridges or road embankments, and destruction of streamside vegetation.

Increased sediment movement from hillslopes may come directly from unpaved road surfaces that are not protected by vegetation. When rainwater or melting snow comes in contact with these compacted surfaces it flows overland rapidly, rather than infiltrating, an effect documented in mountains from the Appalachians to the Cascades.[129] Small surface irregularities cause the water to concentrate along specific flow paths. As water depth increases along these paths, greater force is exerted on the ground surface, enhancing erosion and creating larger

An 1890s view of the road at the mouth of Bear Creek Canyon. (Photograph courtesy of the Colorado Historical Society)

paths, or rills and gullies, which in turn capture more flow in a self-enhancing feedback process. In one 320-acre (132 ha) catchment tributary to the Big Thompson River, erosion of a single unpaved road provided 25% of the basin's sediment.[130] Other studies from the coastal ranges of North America have demonstrated that even a carefully constructed road system may double a basin's sediment production, and that the rate of sediment movement from road surfaces is dependent on the amount of use. Along paved roads, increased sediment movement may result from the sand and gravel commonly dumped on these roads to increase traction during periods of snow and ice buildup.[131]

The mouth of Bear Creek Canyon in June 1995.

The road through Poudre Canyon in 1914. (Photograph courtesy of the Ft. Collins Public Library)

An undated photograph of the railroad along Clear Creek Canyon. (Photograph courtesy of the Denver Public Library, Western History Department)

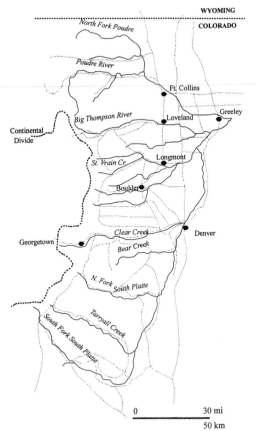

Schematic showing the principal (mostly paved) roads in the Colorado Front Range (light, dotted lines) in relation to the Front Range Rivers (solid lines).

A 1995 view of the maintenance road along the Michigan Ditch in the headwaters of the Poudre River.

A three-year-study along Joe Wright Creek, a tributary of the upper Poudre River, illustrates the effects of increased sediment movement associated with roads. At sites that were influenced by highway construction activities along the stream, suspended solids increased from 10 to 100 times the reference levels. Algal species diversity was reduced, and reductions in density, abundance, and diversity of macroinvertebrates were recorded, as well as changes in species present. Suspended solids returned to reference levels within a year of the cessation of construction, as did macroinvertebrate communities. In this case, the rapid biotic recovery was attributed to colonization from undisturbed areas upstream and in tributary streams. Recovery was aided by the fact that water temperature and flow regimes and water chemistry had not been altered. Moreover, there was minimal deposition of fine sediment in the study reach because of its high channel-bed slope (3%).[132]

Changes in slope stability associated with roads may result from removal of lateral support when a portion of a steeply inclined slope is removed for a roadway. The excavated material is often dumped downslope of the road, and the roadbed may be strengthened with crushed rock or paving materials. The increase and redistribution of mass asso-

Gully forming beside an unpaved road in the Poudre River drainage basin.

Gullies forming along an unpaved road in the Poudre River drainage basin.

Sand spread during the winter months to improve vehicle traction on a paved road in the Clear Creek drainage basin accumulates along a small channel beside the road during spring snowmelt.

Slope failure along a road in the Clear Creek drainage basin. This failure was promoted by the removal of lateral support during road construction.

Slope failure caused by a change in permeability associated with a paved road in the Clear Creek basin.

ciated with these activities may also alter slope stability. Once traffic begins to move along a road, vibrations from the traffic may weaken the internal structure of unconsolidated slope materials. Finally, hillslopes may be affected by changes in subsurface permeability and the downslope movement of water associated with roads. If a hillslope provides a downward path for infiltrating rainwater and snowmelt to reach the valley bottom, then a road is an interruption of that path. Mass is transferred and permeability is decreased. A summary of four studies conducted in the Pacific Northwest noted that the rate of slope failure in forested drainages was increased by up to 4 times by clearcutting, and from 25 to 340 times by road building. Landslides occurring within 280 feet (85 m) of a road in the mountains of Puerto Rico were five times more frequent than those occurring at greater distances from roads.[133]

Changes in river morphology may occur along narrow portions of the valley where road embankments or bridges impinge on the river. The narrow reaches of Poudre and Big Thompson Canyons provide good examples. The river channels historically occupied the entire valley bottom in these reaches, and the first roads avoided these canyon segments. When roads were eventually built along the narrow canyon reaches, the rivers were constricted. A constriction, particularly if it is stabilized by

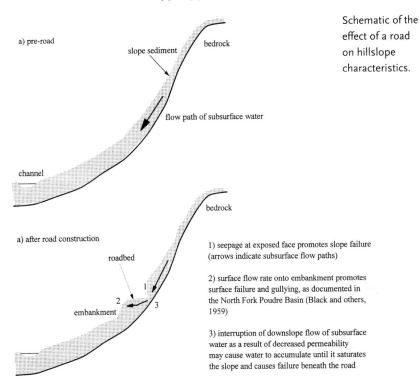

Schematic of the effect of a road on hillslope characteristics.

a) pre-road

slope sediment

bedrock

flow path of subsurface water

channel

a) after road construction

roadbed

embankment

bedrock

1) seepage at exposed face promotes slope failure (arrows indicate subsurface flow paths)

2) surface flow rate onto embankment promotes surface failure and gullying, as documented in the North Fork Poudre Basin (Black and others, 1959)

3) interruption of downslope flow of subsurface water as a result of decreased permeability may cause water to accumulate until it saturates the slope and causes failure beneath the road

bedrock or riprap, results in faster and deeper flow. This may enhance erosion of the channel bed in the narrow reach, with associated deposition in the wider reaches upstream and downstream from the constriction. Similarly, bridges often act as channel constrictions and alter the patterns of erosion and deposition along the bed and banks of the river.

Roads may impact or destroy streamside vegetation, both directly and indirectly, by facilitating recreational uses. Tourists have been coming to the Front Range since the advent of mining in 1859; there was a summer resort at Cherokee Park in the North Fork Poudre River by 1883, for example, and President Theodore Roosevelt was a guest at the Rustic Hotel (1882) along the Poudre River.[134] Fishermen, campers, picnickers, and off-road vehicles all decrease ground-cover vegetation through trampling and decrease the total number of vegetative species.[135] Foot and vehicle traffic increase mechanical injury to mature trees, eliminate seedlings and young saplings, and increase surface compaction. When the surface becomes more compacted, rainwater and snowmelt are more likely to flow quickly over the ground surface rather than infiltrating.

The mouth of Big Thompson Canyon, looking downstream, circa 1914. Note the narrow, unpaved road to the right of the river. (Photograph courtesy of the Ft. Collins Public Library)

The same view of Big Thompson Canyon, taken in the 1930s. Note that the road appears to be more heavily used, but not substantially wider. (Photograph courtesy of the Ft. Collins Public Library)

The view of Big Thompson Canyon in the 1940s. The road has been paved and increased to two lanes with shoulders, substantially decreasing river width. (Photograph courtesy of the Ft. Collins Public Library)

This in turn promotes sheet erosion, truncation of the soil profiles, and root exposure, thereby introducing more sediment to stream channels.

In general, the increase of sediment entering rivers from the valley slopes has probably been the most important consequence of roads built in the Front Range. Few studies have examined these processes in the Front Range, but investigations conducted in similar terrain suggest a substantial increase in sediment movement associated with roads. This would in turn adversely affect channel habitat diversity—by filling pools, for example—and aquatic biota, as described previously. For stream channels within approximately a quarter mile of a road, road-related channel constriction and increased sediment, along with flow regulation, are probably the dominant contemporary physical impacts on the channel.

Recreation

In addition to the road-related aspects of recreation, the biota of the Front Range rivers have been affected by deliberate manipulation of fish communities. As early as 1874, travelers to the Colorado Rockies were noting the excellent trout fishing.[136] The combination of more intense fishing with increasing human population, and the channel changes described

Recreational use along Boulder Creek on a summer weekend. The pedestrian and bicycle path at left is heavily used, as is the road at right. Both constrict the river.

Celebrants of "Good Roads Day" at Indian Meadows along the upper Poudre River, October 12, 1920. The river is at far left rear. (Photograph courtesy of the Ft. Collins Public Library)

throughout this chapter, led to the stocking of hatchery-raised fish in the Front Range rivers. As early as 1909 several University of Colorado professors formed an association to buy a 75-acre (30 ha) ranch in North St. Vrain Canyon to use as a fishing retreat, stocking the stream with trout.[137] By 1970 the Poudre River had a resident cold-water fishery of approximately 810 trout at least 15 inches (140 mm) in length per half mile of river, and the Colorado Division of Wildlife annually stocked approximately 1,050 "catchable" rainbow trout per half mile.[138] By 1994, all of the trout species currently present along the Poudre River, with the exception of the greenback cutthroat trout in tributary headwaters, were introduced. The Poudre River has essentially become a factory for fishing. During May–August 1994, for example, the Colorado Division of Wildlife stocked 20,967 small rainbow trout 2–4 inches (5–10 cm) in length and 15,000 "catchable" rainbows 10 inches (25 cm) in length along a 3.4-mile (5.7 km) portion of the river. Each episode of stocking averaged 3,750 fish, which were removed by fishermen within 7–10 days. Not all of the Front Range rivers are as heavily stocked. The May–August 1994 figures for a lightly stocked 4-mile (7 km) reach of North St. Vrain Creek were 7,600 fish.[139]

Studies of the biotic effects of these stocking programs have not been conducted for the Front Range rivers, but studies on other Rocky Mountain rivers have demonstrated that the stocking of nonnative fish and fish-food organisms such as freshwater shrimp changes the composition and abundance of the native fish species, as well as of their prey organisms.[140]

Grazing

The grasslands of the Colorado Front Range have been grazed by bison, elk, deer, and bighorn sheep for thousands of years, although the numbers and distribution of these animals have been substantially altered during the past two centuries. An estimated 30 million bison roamed North America in 1800; by 1889 the number had fallen to 1,000 and has subsequently risen to 200,000.[141] Despite the historically high numbers of grazing animals, these creatures probably had only temporary, localized impacts on stream channels because they were free to keep moving as they grazed, rather than being confined to a restricted area. Only since the advent of restrictions on both wild and domestic grazing

Incision along Beaver Brook in Beaver Meadows, Rocky Mountain National Park. The river presently has almost no riparian vegetation.

has the presence of these animals been likely to alter stream channels significantly.

In some areas the population of wild grazing animals has increased because of human manipulation. Between 1875 and 1877, elk were nearly eliminated in the Estes Park area of the upper Big Thompson River drainage as a result of hunting. As the region became a tourist attraction, and then a national park in 1915, efforts were made to restore natural conditions. Forty-nine elk were transplanted from Yellowstone to Rocky Mountain National Park during 1913–14, and hunting was banned. By 1931 it was noted that the rapidly growing elk herd was damaging aspen bark in an effort to find winter forage. In 1944 selective hunting was allowed, but this practice was discontinued in 1968 under the assumption that "natural self-regulation" and hunting outside of the park boundaries would control elk numbers. However, the elk herd has continued to grow in size. Elk prefer wet meadows and aspen for their winter range, and soil compaction and channel erosion in areas such as Beaver Meadows have been attributed to overgrazing by elk.[142]

Domestic livestock and draft animals were introduced to the Front Range by the early Euro-American settlers, but the lack of suitable graz-

A grazed reach of Sheep Creek, a tributary of the Poudre River. Browsing cattle have eliminated shrubby riparian vegetation and trampled the banks, creating a wider and shallower river channel.

ing land kept the numbers of animals relatively low.[143] In 1884 there were only four ranches in the Clear Creek valley, and these were restricted to a strip approximately 2 miles (3 km) long on the south side of Clear Creek. Many of the open meadows of the Front Range did receive locally heavy grazing pressure. The Estes family drove cattle into Estes Park in 1860. By 1900 local ranchers perceived cattle grazing in the park to be excessive, but in 1939 grass still grew in the park only where the land was fenced to keep out cattle. Similarly, heavy grazing along Sheep Creek, below Eaton Reservoir in the North Fork Poudre basin, has enhanced stream bank erosion. The regulation of grazing lands did not begin until 1904.

Because grazing along the Front Range rivers is limited to fairly small meadow reaches, it has not had nearly as substantial an effect on these rivers as have other land uses such as mining. However, overgrazing has had dramatic impacts on limited reaches of these rivers.[144] Nearly all of the meadows along the perennial channels have been heavily grazed by cattle and other livestock. Approximately 300,000 cattle grazed the region in 1940, although the number was reduced to 120,000 by 1983. Livestock prefer streamside areas because of ample green forage, water supply, and cooler temperatures. A study along the South Platte tributary Trout Creek found that cattle spent about 5% of each day in the channel, and more than 65% of each day within 300 feet (100 m) of the stream.

The effects of overgrazing on the entire river corridor have been documented in case studies for mountain rivers throughout the western United States.[145] Grazing animals compact the soil, which increases surface runoff at the expense of infiltration, thereby accelerating erosion, decreasing water quality, and causing fine sediment to accumulate on the channel bed. Trampling also often reduces bank stability, causing wider and shallower channel cross sections with less of the undercut banks that are used by fish. Wider and shallower rivers generally have higher summer water temperatures, an effect that is enhanced when grazing animals eliminate streamside shrubs and decrease vegetative cover for the river.

Streamside vegetation provides cover for trout by creating quiet, shaded resting areas beneath overhanging vegetation, and by contributing material to debris jams. The roots of streamside plants are critical to the development and maintenance of undercut banks, which also provide cover for trout.[146] The roots also help to stabilize channel banks, thus reducing siltation in pools and on spawning bars. Comparing fish habitat along a fenced reach of 1.5 miles (2.5 km) of Sheep Creek with habitat along grazed reaches, one study found that the fenced reaches were narrower and deeper, had less streambank alteration and better streamside vegetation, and an estimated trout population two times that of grazed reaches.[147] Livestock grazing has been identified as the single greatest threat to the integrity of trout-stream habitat in the western United States.[148] This activity has historically been intense and widespread in mountain regions as diverse as the Himalaya, the European Alps, and the Sierra Nevada.[149]

Animal wastes excreted in or near streams add nitrogen to the river ecosystem.[150] Nitrogen is an essential nutrient that cycles through aquatic and terrestrial environments in various forms. The river corridor is particularly important for processing nitrate-nitrogen from upslope ecosystems; bacteria and blue-green algae occurring in this corridor mediate the processes that convert nitrogen into its various forms such as nitrate or ammonia, and the river corridor may act as a source or sink for nitrate-nitrogen. If heavy concentrations of grazing animals substantially alter the nitrogen input, the corridor's buffering role may

(a)

(b)

Adjacent grazed (a) and ungrazed (b) reaches of a small river in the Poudre River drainage basin. The ungrazed portion has dense riparian vegetation with several edges and strata.

Looking downstream along an ungrazed reach of Sheep Creek. Overhanging shrubby willows provide shaded resting areas along the river banks, and the river is relatively deep and narrow.

be overwhelmed. When nitrogen-rich pollutants reach a channel they act as fertilizers that may encourage algal blooms, kill fish, and cause eutrophication of the river. In addition, cattle excretions may cause water in a river to fall below bacteriological standards set by public health agencies. For the State of Colorado these standards specify that less than 200 fecal coliforms/100 milliliters (or 0.1 quarts) can be present for recreation class I—primary contact—water. A single cow daily produces and eliminates an estimated 5.4 billion fecal coliforms and 31 billion fecal streptococci. A study along Trout Creek found that about 5% of the total manure produced by cows contributes to stream channel pollution, and that moderate grazing does not significantly affect water quality.[151]

Many forms of wildlife rely on river corridors and are thus adversely affected by grazing. Examples from the Rocky Mountains include a study demonstrating that the ungrazed portion of a stream had more garter snakes than did another portion of the same stream as a result of having more vegetation and organic debris.[152] Similarly, the presence and abundance of many species of small mammals including mice, voles, shrews, and chipmunks, all of which have a differentially high use of streamside habitat, are adversely affected by disturbances such as grazing.[153]

The effects of livestock grazing vary greatly as a function of intensity and duration of grazing, and of channel type. Fortunately, many stream-side areas show substantial recovery of pregrazing conditions within 1–2 years of their removal from grazing, although the rate of recovery varies as a function of channel type. It is often economically more efficient to reduce or eliminate grazing than to install fish habitat structures when trying to restore a river's fish community.[154]

The End Result?

The rivers of the Colorado Front Range have obviously been altered to varying degrees in association with several land-use activities since 1859. In general, the response of any system to an external change will be a function of the magnitude, frequency, and duration of that external change. Both the physical and biotic characteristics of rivers seem to return to predisturbance levels within a decade following the cessation of activities such as overgrazing, whereas rivers used for railroad-tie drives remain demonstrably different than unaffected rivers a century after the tie drives occurred. It is difficult to make a definitive assessment of river changes during the period 1859–1990 because there are so few baseline data. Pre-1859 records of the Front Range rivers are sparse and qualitative, and only a few of the smaller tributary streams have not been affected by some type of land use. These relatively unaltered rivers are of inestimable value because they provide analogs or baselines against which to compare altered rivers, and because they provide biotic reservoirs from which organisms can potentially migrate to colonize disturbed channel reaches once the disturbance ceases.

Perhaps most remarkable about the Front Range rivers is their ability to recover from multiple, prolonged disturbances. Tenmile Creek west of Denver provides a spectacular example of this resilience.[155] During the 1880s two narrow-gauge railroads each changed the creek's location. The railroads were followed during the twentieth century by a gas pipe-line, several power distribution systems, and a buried telephone cable, each of which had associated roads and construction. Upstream mining released toxic drainage and mill tailings that covered fish spawning beds, and during the 1970s 3 miles (5 km) of channel were relocated to facilitate the construction of Interstate Highway 70. During just two months of a 1976 study along the creek, four spills occurred directly into

the creek: 17 September-a "large quantity" of silt from a mining operation; 9 November—between 120,000 and 250,000 gallons (32,000–
66,000 L) of mill tailings, which included zinc and arsenic, from a
broken 4-inch diameter (10 cm) pipe at the same mining operation;
17 November—approximately 5,000 gallons (1,325 L) of gasoline from
an overturned transport; 18 November—between 500,000 and 750,000
gallons (132,000–198,000 L) of tailings from the same mining operation. Despite these occurrences, the creek still supported fish; creek relocation efforts included $.5 million for providing fish habitat equal to
that prior to construction. The relocated channel serves as an important
reminder that we are not yet competent to restore rivers. At Tenmile
Creek, log and rock check dams, log and rock deflectors, and large boulders were used to stabilize the new channel and to provide habitat diversity. All of these structures worked reasonably well for about a year
and a half, until a large flood removed most of the check dams, particularly in steep channel reaches. To quote the scientist summarizing the
results of the channel relocation: "The . . . flood contoured the stream
components much like the original channel regardless of man's efforts."

The presence of aquatic biota and functioning rivers in the Front
Range should not be interpreted as an indication that land-use impacts
on these rivers are inconsequential. Only the most severely degraded
rivers will be biologically dead. As noted throughout this chapter, the
usual effect of river disturbance is to reduce species diversity by reducing habitat diversity and favoring those species that are resistant to
frequent disturbance or to pollutants. A segment of river that appears
scenic and attractive to the untrained observer may in fact be a physically
simplified, biologically impoverished remnant of the river that existed
prior to the nineteenth century. Woody debris and a well-developed pool-
riffle sequence may be absent, allowing coarse sediment to move rapidly
and frequently through the reach and creating a fairly uniform habitat
that supports a limited number of species. Recolonization of the reach
by aquatic biota may be inhibited by the high sediment mobility and
low habitat diversity, as well as by repeated introduction of pollutants
ranging from slope sediment to cattle manure and heavy metals. Although aquatic biota are present throughout the Front Range rivers,
their abundance and diversity have undoubtedly been reduced by multiple land-use-related impacts during the past two centuries.

The physical responses of the Front Range rivers to various distur-

A pedestrian and bicycle path along Boulder Creek intersects a canal pipe (center) and parallels an abandoned roadbed (left). The modern road was blasted straight through the bedrock ridge at the extreme left of the photograph.

A view downstream along the Clear Creek valley. The creek (left) is narrowly confined as it flows through Georgetown. Downstream, flow is impounded in a reservoir (center rear). The embankment of Interstate 70 (far left) also encroaches on the river. Abandoned tailing piles form light patches at the right rear.

bances have probably been constrained by the relatively resistant channel boundaries. The bedrock and, to a lesser extent, the coarse sediment in which these rivers are formed have probably reduced the magnitude of channel filling, scouring, and lateral mobility resulting from changes in the ratio of water and sediment entering the rivers. The downstream variability caused by glacial deposits or by changes in the underlying geologic structure and rock type have probably also constrained river response. For example, channel erosion caused by removal of resident beaver along a low-gradient meandering river reach would be unlikely to continue upstream through a steeper reach where cobbles and boulders formed steps and pools.

The Front Range rivers appear relatively pristine today partly because of these "buffering" effects, and partly because most people do not know how the rivers might have changed during the past 200 years. The pristine appearance may be misleading, but it is a quality that our society increasingly values. Many of the contemporary issues in water management strive to balance this public appreciation for natural rivers against increasing societal demands for water, recreation, and living space.

Chapter 4

The Present, and the Future

The Colorado Front Range urban corridor presently has a population of almost 3 million people. Demographic projections to the year 2020 predict a population of 3.8 million.[1] Many of these people have moved, or will move, to the region at least in part because of its outdoor recreational opportunities. The Colorado Rockies are famous for skiing, camping, hiking, fishing, mountaineering, and whitewater boating. Annual visitor statistics for Rocky Mountain National Park, in the heart of the Front Range, vividly document the rapid growth in recreational use of the region. In 1976 the 380-square-mile (1,070 km²) national park issued 53,943 backcountry permits. In that same year, there were 308,057 overnight stays in the park's five campgrounds, and 500–700 cars per hour on the park's main thoroughfare, Trail Ridge Road. These numbers are not suprising in view of the fact that the populations of nearby Boulder and Larimer Counties grew to 166,500 (+26.2%) and 120,900 (+34.5%), respectively, between 1970 and 1975.[2] Camping in Rocky Mountain National Park increased steadily during the 1980s and 1990s (Bonnie White, Rocky Mountain National Park, pers. comm., 1997). Regional population has also continued to increase steadily. By 1995 the populations of Boulder and Larimer Counties were at 255,050 and 215,680(increases of 53% and 78% relative to 1976), respectively. It is in this context of rapid growth in human population and recreational use that it is necessary to consider the future of the Front Range rivers.

Population and recreational pressures, along with other human impacts on the Front Range rivers, are partially controlled by existing legislative protection. Some of the earliest such legislation dealt with reservation of public lands and limitations on land use. A substantial portion of the upper South Platte River basin is reserved as national forest or national park land. The degree of protection associated with these land

Table 4.1. Levels of recreational activity in the Colorado Front Range

Rocky Mountain National Park		
Year	Number of visitors	Ratio of increase relative to 1915
1915	31,000	—
1948	1,032,000	33
1968	2,187,000	71
1972	2,519,622	81
1978	3,037,000	98
1994	3,153,695	102

Estes-Poudre Ranger District, Roosevelt National Forest[a]						
Year	Camping	Picnicking	Trail use	Canoe/Raft	Fishing	Scenic travel
1985	45,700	9,000	29,500	1,500	34,300	575,000
1990	40,900	10,600	48,200	2,800	38,000	612,600
1993	59,200	10,100	55,000	16,700	38,000	600,000
2000	71,000	11,100	77,300	30,000	41,200	621,300
	(1.6)	(1.2)	(2.6)	(20)	(1.2)	(1.1)

Source: Information for Roosevelt National Forest from Gordon Hain, U.S. Forest Service, Roosevelt National Forest, 17 Feb. 1995, personal communication. Numbers for the year 2000 are projected.
[a] 12-hour recreational days (ratio of increase relative to 1985)

designations varies. Drainage basins in Rocky Mountain National Park are subject to moderate recreational and road-related impacts and in some cases to overgrazing associated with uncontrolled elk populations. Water diversions and small irrigation reservoirs built before the park was established in 1915 are still in use, but these structures are gradually being acquired and dismantled by the Park Service. National forest lands in the Front Range are presently used primarily for recreation rather than timber harvest. The drainages on these forest lands have many water diversions and are subject to future diversions. The drainages also experience heavier recreational and road impacts than drainages on national park lands, but they are less affected by urbanization than adjacent private lands. In general, the portions of the Front Range rivers on public lands are less likely to be affected by human activities during the next few decades than are rivers on private and commercial lands.

Other types of legislation have also been very important in determining the conditions of mountain rivers in the Front Range. As early

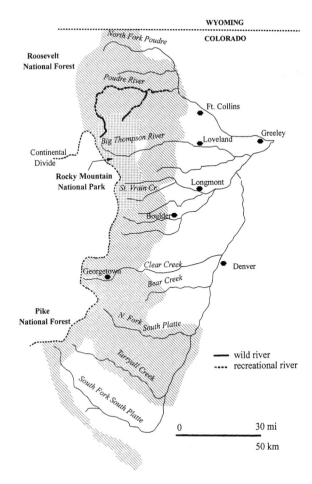

Schematic map of national park (stippled area) and national forest (diagonal lines) lands in the upper South Platte River basin. Federally designated reaches of wild river and of recreational river along the Poudre River are also shown by solid and dotted lines, respectively. The portion of the Big Thompson River flowing through Rocky Mountain National Park receives similar protection.

as 1960, the National Park Service recommended preservation of free-flowing rivers,[3] and in 1963–64 the secretaries of interior and agriculture oversaw the Wild Rivers Study that served as the basis for the initial wild rivers proposal. Rivers designated under the 1968 Wild and Scenic Rivers Act were classified as wild, scenic, or recreational, depending on the land-use impacts along each river at the time of its designation.

Because of potential conflicts with water-development advocates, Representative Wayne Aspinall (R-Colo.) played a major role in ensuring that no rivers in Colorado were included as part of the system initially. However, numerous amendments to the original act have expanded the number of designated rivers. A 1975 amendment dealt with twelve rivers in Colorado, including the Poudre River.[4] In 1986, approximately 75

A portion of the Poudre River designated as a national recreational river under the Wild and Scenic Rivers Act.

miles (125 km) of the Poudre River were designated as Colorado's first Wild and Scenic River. Thirty-one miles (52 km) of the upper portion of the Poudre, and lower portion of the South Fork Poudre, were designated as a wild river that is essentially primitive and unpolluted, free of impoundments, and generally accessible only by trail. Forty-four miles (73 km) of the central portion of the Poudre, and segments of the South Fork Poudre, were designated as a recreational river that has some impoundments and development, and is readily accessible by rail or car. As with the protection accorded by public land status, these portions of the Poudre River basin are likely to be less impacted by future human activities than are adjacent, unprotected rivers.

The remaining principal channels of the Colorado Front Range have not received similar wild and scenic designations because they are perceived as being too altered by human activities. Paved roads run adjacent to at least a substantial portion of the length of the channels of the Poudre, Big Thompson, St. Vrain, Boulder, Bear, Clear, North Fork South Platte, Tarryall, and South Fork South Platte, and many of these roads receive heavy traffic. There are at least 47 towns of 1,000 or fewer people, and five towns of 1,000–5,000 within the upper South Platte River basin,

Table 4.2. Volume of vehicular traffic along selected Front Range
roads

Road segment	Length (km)	Annual average daily traffic, 1993[a]	Predicted volume, 2013[b]
State route 34 (Big Thompson)			
Estes Park	1.9	14,000	18,200
Narrows	2.5	4,050	4,860
Interstate 70 (Clear Creek)			
Loveland Pass	16.0	22,800	36,480
Idaho Springs	5.4	30,800	46,200
State route 14 (Poudre River)			
Upper canyon (Eggers)	15.2	780	1,560
Lower canyon (county road 29C)	1.9	1,300	2,600
Mouth of canyon (state route 287)	0.3	11,500	27,600

Source: Colorado Department of Transportation, Division of Transportation Development, Information Management Branch
[a] Total number of vehicles traveling in both directions for a year, divided by 365
[b] Straight-line projection of historical records of traffic volume

with more than a dozen cities along the base of the mountains, including eight communities with populations greater than 50,000.[5] On the one hand, people expect the Front Range rivers to at least have the appearance of natural mountain streams, complete with catchable trout. On the other hand, residents of the Front Range and the neighboring plains expect to have abundant water for agricultural and municipal uses. This water is largely taken directly from the Front Range rivers, or it is delivered from the western slope via the Front Range rivers, thus creating a basic dichotomy between recreational/esthetic and utilitarian expectations. The concept of instream flow has been developed as a means of mediating between these contrasting expectations. Instream flow is at the heart of the first of the key issues for the mountain rivers of the upper South Platte River basin in the twenty-first century: water quantity.

Instream Flow

In the simplest sense, the phrase "instream flow" refers to the water actually flowing within a stream channel. Instream flow may refer specifically to a minimum volume of flow, or to a carefully detailed flow regime for which volume and timing are specified. In either case, flow

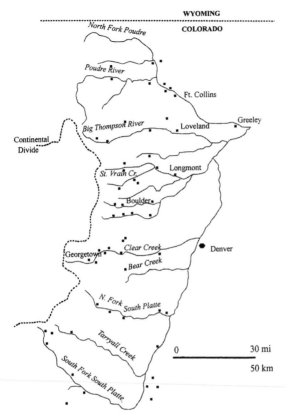

Schematic of urban communities in the Colorado Front Range. Symbols do not indicate relative size of communities. Communities along the eastern base of the Range are not all shown.

characteristics are specified to preserve a desired attribute of the stream channel, such as fish habitat, streamside vegetation,[6] or physical channel capacity to carry runoff. Without legally guaranteed levels of instream flow, some of the mountain rivers of the Colorado Front Range are, or will be, literally drained dry.

Instream flow has become increasingly important as concern has grown over endangered species, and as the regional economy of the Front Range has changed.[7] With the economic decline of agriculture, mining, and energy industries during the 1980s and 1990s, recreation and tourism, and the wild and scenic rivers that support them, have correspondingly increased in economic importance. For example, then-Governor Richard Lamm noted in a 1985 speech that alfalfa, which consumed 27% of Colorado's water, annually added $156 million to the state's economy. Recreation and tourism had an annual contribution of $4 billion.

(a)

(b)
Two contrasting scenes along the Poudre River in 1995: (a) upstream view of a diversion structure that removes flow from the river and creates an artificial drop; (b) recreational whitewater rafters.

The protection of instream flow is a complicated and expensive process set about with many legal requirements.[8] Legal precedents related to instream flow date back to three pieces of nineteenth century legislation: the 1866 Mining Act, an 1870 amendment to the act, and the 1877 Desert Lands Act. These acts have been interpreted by the U.S. Supreme Court as having given states, rather than the federal government, the authority to determine the method by which the right to use water would be allocated on the public domain. In Colorado, this authority is held by the courts, the state engineer, and the Colorado Water Conservation Board (CWCB). The 1876 Colorado State Constitution holds that "the right to divert the unappropriated waters of any natural stream to beneficial uses shall never be denied." However, in 1979 the Colorado Supreme Court held that this statement was a rejection of riparian rights doctrine, which gives water rights to lands adjacent to a river channel, and not a requirement that a water right has to involve diversion. This 1979 ruling provided the opportunity to create instream flow water rights. In 1973 the Colorado legislature requested that the CWCB establish water rights on behalf of the public to "preserve the natural environment to a reasonable degree." Although only the CWCB can apply for or hold an instream water right, individuals can acquire and donate instream rights, and state and federal agencies such as the Colorado Division of Wildlife or the U.S. Departments of Agriculture or Interior can request such rights.

The process of purchasing or requesting instream water rights is complicated and expensive. In the case of public lands, the ability to acquire these rights depends largely on the courts' interpretation of the legislation originally authorizing the reservation of public lands. The Colorado Supreme Court in 1982 upheld reserved water rights for Rocky Mountain National Park because of the court's interpretation of the original intent of Congress in establishing national parks.[9] In the case of national forest lands, the court ruled in 1978 that although the federal government possessed federal reserved water rights to the use of water in the national forests, those rights were limited and did not include instream flow for maintaining river ecosystems. As a result, the Forest Service has tried to argue for flows based on channel maintenance.

Channel-maintenance flow refers to the discharge necessary to maintain such aspects of river form as width/depth ratio or pools and riffles.[10] One of the difficulties in designating a channel-maintenance flow involves deciding which aspect of channel morphology is of interest, because different magnitudes and frequencies of flow are responsible for maintaining various portions of the channel.[11] For example, the boulders that form riffles in high-gradient channels may become mobile at only the highest discharges; large floods are thus necessary to create and maintain the pool-riffle sequence. But it may be the moderate-sized flows that flush fine sediment from the riffles and bars, and thus maintain the microhabitats necessary for fish spawning.

Often, bankfull discharge is assumed to be the most important flow in controlling channel morphology.[12] Most precisely, bankfull discharge is the discharge that fills the channel to the top of the banks. The term is also sometimes used to describe a discharge that occurs on average once every 1–2 years. This second usage assumes that the bankfull discharge is equivalent to the mean annual flood, which may or may not be true. Analysis of discharge and sediment transport records for rivers across the United States indicated that the fairly frequent, bankfull floods controlled channel characteristics in many alluvial rivers. However, for desert rivers with extremely variable hydrologic regimes, or for mountain rivers with resistant channel boundaries formed by bedrock or coarse sediment, it may be only the infrequent, large floods that are capable of mobilizing a substantial quantity of sediment and thus altering channel form.[13]

In 1990, the U.S. Department of Justice, representing the Forest Service and acting on behalf of the United States, filed federal reserved water right claims for channel-maintenance flows in the Colorado Water Division 1 Trial.[14] These claims to instream flows were challenged by the State of Colorado and by water conservancy districts in northern Colorado that divert water from national forests. In February 1993 the presiding judge ruled that although the reserved water rights of the United States include channel-maintenance purposes, the United States had failed to demonstrate that the reserved water rights claimed were necessary to preserve the timber and water characteristics which form the purpose for reservation of the national forests. Technical testimony in the case largely focused on whether bankfull discharge was necessary to maintain the mountain river channels. The United States argued that bankfull discharges every 1 to 2 years were necessary; the State argued that the channels were relatively unaffected by bankfull discharge, being instead shaped by the higher discharges present during melting of the Pleistocene glaciers, and by rare, large floods such as the 1976

Big Thompson flood. Both sides in the case were hampered by the limited data on channel dynamics available for the mountain rivers in question, and the U.S. Forest Service embarked on an extensive program of field measurements during and after the trial, in preparation for similar court cases in other western states. And, because the legal framework required that instream flow be considered only in relation to channel form, no attention was given to the role of water diversions in impacting river function and impoverishing the river ecosystem, effects that have been demonstrated by numerous studies in other mountain river environments with altered flow.[15]

Acquisition of instream flow rights by individuals or nonprofit groups may be equally difficult. Colorado does not recognize recreation as a valid purpose for seeking an instream flow water right,[16] despite the importance of water-based recreation to the state economy. As a result, individuals seeking to acquire instream water rights must base their arguments on protection of the natural environment. This may involve extensive data acquisition and legal fees, as well as the purchase cost of existing water rights. Despite these complications, as of 1989 the cwcb had established new water rights on more than 6,600 miles (11,000 km) of rivers in Colorado. These water rights typically designate a specified flow level within a river segment which is, on average, 7 miles (11.5 km) in length[17] during each season. However, holders of senior water rights are still allowed to deplete the stream below the specified flow level.

Another potential means of preserving instream flow rights is that of federal environmental legislation. The U.S. Constitution supremacy clause gives precedence to federal laws if the federal laws are in conflict with state laws. Thus, the Endangered Species Act, the Clean Water Act, the Wild and Scenic Rivers Act, and other pieces of federal legislation may be used to protect instream flow if it can be demonstrated that instream flow is necessary to an endangered species, to clean water, or to some other commodity protected by the legislation. Even in these cases, however, instream flow may not be used to deny existing diversions, but only to deny proposed new offstream water uses.

Instream flows provide relatively environmentally benign economic benefits by supporting fishing, whitewater rafting, hunting, birdwatching, and camping. The flows may also save communities millions of dollars in sewage and effluent treatment costs by providing dilution that helps to ensure that receiving waters do not exceed contaminant standards. Despite these benefits, the key question in the water-limited Colorado Front Range becomes, what is a sufficient minimum instream flow to guarantee protection of a specific resource? Because of the legal neglect of recreation, and legal acceptance of environmental protection as a basis for instream water rights in Colorado, instream flow levels are most often based on the habitat needs of fish, although not necessarily native species. The difficulty arises in deciding how to assess the habitat requirements of individual fish species. For example, is the aim to ensure survival, or to create optimum species production? And what, exactly, does an individual fish species require? Presumably each species is adapted to the natural flow regime of the rivers in which it historically was present. However, if water is being transferred into and out of the river, that flow regime is being disrupted. How much disruption can the fish, and the other organisms on which they depend, tolerate?

Following enactment of the National Environmental Policy Act in 1980, increasing attention was given to the evaluation of various alternative designs and operations of federally funded water projects. This required methods capable of evaluating the environmental impacts of a series of possible alternative development schemes,[18] usually through modeling the effects of varying levels of instream flow.

The modeling approach to instream flow specification is based on quantifying the interacting variables that make life possible for a given fish species.[19] The approach begins with Habitat Suitability Index models, such as those developed by the U.S. Fish and Wildlife Service. These models include an overview of the designated species' life history and known habitat requirements by life stage. The idea is to have an objective, quantifiable method of assessing existing habitat conditions for the species within a study area. As a hypothetical example, brown trout cannot reproduce in a silty substrate, but females manage to produce a few viable eggs in a substrate of sand and leave a bumper crop in sandy gravel. In a Habitat Suitability Index model for brown trout eggs, silt would receive a rating of 0.0 (unsuitable), sand might rate 0.5, and sandy gravel would be 1.0 (optimal).

The next step is to use a model such as the U.S. Fish and Wildlife Service's Instream Flow Incremental Methodology to assess the effects of changes in water flow on habitat of the species by life stage.[20] In other words, how much of that optimal sandy gravel is available for spawning if discharge is reduced by 10% during the summer? Finally, the models may be nested with models of hydrologic and hydraulic variables, as well as models of reservoir operations or irrigation demands, to create basin-wide methods of instream flow management.[21]

Developing the model components for a single species on a particular river segment may require two years of analysis.[22] Obviously, the models are only as good as our understanding of both the hydraulic and sediment transport processes occurring in rivers, and of the habitat needs of individual species throughout their life history. One of the potential shortcomings of models or approaches that focus solely on a constant minimum flow is that annual or seasonal flow variability may be a key control of biodiversity in many rivers.

Criticisms of instream flow models have focused on their simplifications of a complex reality, as summarized in critiques by Robert J. Behnke.[23] The assumption underlying most instream flow models is that if flow falls below a minimum level, a decline will occur in the target species. However, direct cause-and-effect relations between flow-related habitat and fish have not been quantitatively verified. During the past 30 years, a variety of techniques have been used by state and federal agencies to make flow recommendations, but none of these techniques have demonstrated an ability to quantify changes in aquatic values with changes in flow. On the contrary, numerous studies document the failure of model predictions to correlate with the abundance or biomass of a target species.

The primary problem with instream flow models seems to be that changes in physical habitat do not provide a basis to accurately predict changes in the target species. There are two reasons for this. First, accurate predictions based on observations from present and past conditions are possible only if the system under observation is mostly stable, isolated, and capable of returning to its original condition following a disturbance. Such systems are extremely rare in nature, where the complexity demonstrated by apparently highly stochastic fish populations is more typical. Second, there are numerous factors influencing a species' well-being, including flow regime, habitat quality, water quality, food abundance and availability, predation, competition, movement, and migration. A habitat model based on limited factors, such as flow depth and velocity, is thus necessarily constrained in accuracy. For example, species A may never occur where species B is present, no matter how good the habitat is for A. Because community assemblages in rivers are constantly changing in response to changes in physical habitat and in the interactions among species, the instream flow models may not accurately predict the consequences of changes in discharge.

Despite these limitations, these models remain the most widely used approach to quantifying instream flows. If scientific understanding alone was at stake, the poor predictions of instream flow models would simply be a spur to further research designed to enhance our understanding. But in the legal arena, the poor predictive ability of the models makes it very difficult for instream flow proponents to argue effectively for a specific discharge scenario. The problem is that, in a semiarid region with increasing human population, increasing demand for consumptive water use places the burden of proof on agencies or organizations seeking to protect river ecosystems. Someone who wants to divert water from a channel is not responsible for demonstrating that the proposed diversion will not alter river form or function. Instead, this responsibility falls on those who seek to limit the diversion. Because any limitation on water use is likely to be challenged in court, it becomes necessary to quantitatively predict river changes resulting from diversion and, if the instream water right is granted, to carefully tailor the timing, magnitude, and location of flows to meet the specific needs of a chosen species or community. The necessity to predict the physical and biological impacts of changes in flow along the Front Range rivers is likely to increase in the future.

Urbanization

Any scientist working on developing channel-maintenance or instream flow models incorporating the needs and responses of aquatic biota will readily affirm that there remain important gaps in our understanding. One of those gaps concerns how the combined stresses of physical habitat loss or flow reduction and reduced water quality may affect aquatic biota. Now that mining has largely ceased within the Front Range, the principal causes of reduced water quality that are likely to grow in importance are recreation and urbanization. Water quality is the second of the critical issues for the mountain rivers of the upper South Platte River basin in the twenty-first century. As noted earlier, recreation reduces water quality primarily through enhanced sediment production. The effects of urbanization are more complex.

Urbanization in mountainous regions is widespread, affecting portions of the Appalachians, the Sierra Nevada,[24] the European Alps, the Himalayas, and the mountains of Japan and southeast Asia, among other areas. Numerous studies conducted across a broad range of environments have demonstrated that, for a given amount of precipitation, the increase of paved areas as a result of urbanization tends to move more water more rapidly into stream channels. This creates larger flows in the river, more frequent moderate floods, lower baseflows, more frequent overbank flows, or an increase in velocity that results in channel erosion.[25]

Urbanization may also increase the movement of sediment into rivers.[26] During the site-preparation phase of construction, for example, the normal, permeable soil cover and protective vegetation are removed. As a result, the subsoil may erode readily, increasing sediment movement into nearby channels. A study conducted in the vicinity of Washington, D.C., found that the average sediment yield of less than 100 tons/mi^2/yr (80–200 tons/km^2/yr) increased to 300–800 tons/mi^2/yr (400 t/km^2/yr) under farming, and up to 100,000 tons/mi^2/yr (55,000 t/km^2/yr) during construction. This increased sediment yield caused deposition of bars in the river; erosion of channel banks as a result of deposition in the channel; obstruction of flow and increased out-of-channel flooding; and unstable channel-bottom configurations. Bottom-dwelling organisms were blanketed with sediment, and stream-community composition was altered as a result of changes in light transmission and the abrasive effects of the sediment, with consequent changes in predatory fish. Sixty percent of the construction sites examined for this study remained open 8–9 months, but once an area is fully urbanized the sediment production decreases drastically, even as water yield is increasing. Streams often undergo erosion as a result of this reversal in the ratio of water to sediment.

Construction may also affect stream channels through runoff contaminated with oils used as machinery lubricants, metals and inorganic chemicals leached or corroded from building materials, and organic waste from litter and food.[27] Other sources of urban contaminants include industry, which may allow leaching from stockpiles of materials, handling of materials, leakage from corroded pipes or storage facilities, and dissolving of pollutants on exposed surfaces. Runoff from streets and highways may be contaminated by residues from rubber, oil, lead

Construction of a parking lot along the North Fork of Clear Creek, June 1995. In this case, no protective fences or other structures have been used to reduce sediment movement from the construction site to the creek.

compounds, asbestos, and asphalt. Residential neighborhoods can produce organics, bacteria from excreta, inorganic nutrients from fertilizers, and toxins from pesticides and herbicides. A 1970–77 baseline study comparing a series of water-quality parameters at a site along the Poudre River above the city of Ft. Collins and at another site 18 miles (30 km) downstream found substantial increases in dissolved solids, turbidity, suspended residue, ortho-phosphate, ammonia, nitrate, and fecal coliforms at the downstream site, as well as decreased levels of dissolved oxygen.[28] Although urbanization has historically been very limited in the Front Range, the increasing population density and willingness of people to commute between scenic homesites in the mountains and jobs in Denver and other major urban centers will make the impacts of urbanization on mountain rivers more widespread in the twenty-first century.

Water Quality and Ecotoxicology

The increase in pollutants reaching stream channels as a result of urbanization and other land uses has prompted increased concern over the cumulative biological effects of these changes. Water quality and biotic response may be assessed in a number of ways. The U.S. Geological Survey's National Water Quality Assessment (NAWQA) program provides an example of a broad, long-term sampling program designed to assess these issues.[29] The Survey has been collecting and interpreting water quality data for more than 100 years, with initial interpretations directed toward the suitability of water for domestic consumption and irrigation. In 1986 Congress authorized the Survey to establish NAWQA, a pilot program to assess the quality of the nation's surface and ground waters. This program is designed to describe water quality conditions, develop hypotheses regarding the major factors influencing water quality, and define data needs. The program has included 60 study units, including the South Platte River basin. The South Platte NAWQA study was completed in 1995. Within the NAWQA study units, there are fixed sampling stations for collecting physical, biological, and chemical constituents. If water quality changes through time as a result of pollutants associated with such activities as urbanization, NAWQA and other sampling programs are designed to be able to quantify the changes.

The major findings of the South Platte River basin NAWQA study are summarized in a U.S. Geological Survey circular.[30] The study indicated that withdrawals of large volumes of water from streams in the basin for agricultural and urban use have resulted in less water to dilute contaminants in the streams. Concentrations of contaminants in channel-bed sediment and whole fish tissue were lowest in forested mountain sites and highest in urban and mixed urban/agricultural areas. The relative abundance of families of fish was altered, and the number of invertebrate taxa was lower, in mining-affected sites and in urban, agricultural, and mixed land-use settings than in minimally impacted areas. Development in mountain drainages correlated with elevated concentrations of dissolved solids, suspended sediment, and nutrients in streams. Alteration of the natural flow regime has degraded native aquatic habitat along streams. The study concluded that surface water quality in undisturbed forested mountain regions was generally good.

Once water quality changes have been identified through programs such as NAWQA, the next step is to relate water quality to the abundance and diversity of aquatic biota.[31] Since the late 1970s, laboratory experiments have exposed various aquatic organisms to contaminated sediments for a period of time to test sediment toxicity. Among the organisms used for such tests are plankton; macrophytes such as duckweed; fish; amphibians; and macroinvertebrates. Each organism has different levels of sensitivity to contaminants, as well as to different types of exposure. Bottom-dwelling organisms may be exposed to contaminants through the ingestion of sediment, from overlying water, and from water within the sediment moving across respiratory surfaces and body walls. Other aquatic species may be exposed to contaminants when burrowing activities of bottom-dwelling invertebrates pump pore-water constituents out of bottom sediment and into the overlying waters, or when an erosive current stirs up the bottom sediment. Other species may also incorporate contaminants from bottom-dwelling prey species.

Increasingly, the effects of contaminants are being studied in the context of ecotoxicology.[32] Whereas toxicology is concerned with the mechanistic bases of the effects of chemicals on individual organisms, ecotoxicology focuses on the effects of environmental contaminants at levels of organization higher than that of the individual, such as populations, communities, and ecosystems. This change in focus is a recognition that no individual organism exists in isolation. If we are concerned about the quality of water necessary to maintain a functioning river ecosystem, then we must acknowledge that the impacts of changing water quality may operate at many levels within the river's food web. For example, contaminants may change the composition of the bottom-dwelling invertebrate community, including species present, as well as the numbers, biomass, life histories, and the spatial distribution of populations. Once the community structure of these invertebrates is altered, these alterations spread to other members of the ecosystem through changes in predator-prey relationships, so that ultimately game fish, raptors, or humans may be affected.

Macroinvertebrates are particularly appropriate for community assessment because their communities are very heterogeneous, increasing the probability that at least some of these organisms will react to a particular change in environmental conditions. These short-lived, quickly reproducing, diverse organisms may provide the first indication of a change in water quality. However, the high spatial heterogeneity of the macroinvertebrate community necessitates much replication in sampling.[33]

The ultimate aim of ecotoxicology, as of any branch of science, is that of using understanding for prediction. Given an environment such as the Front Range, with increasing human impacts, can we realistically predict the effects of alternative development scenarios, for example? One method for attempting such predictions is ecological risk assessment. Ecological risk assessment and management seeks to evaluate the potential of various substances to alter ecosystems. As applied to rivers, this involves four phases. The first phase is that of gathering information on the physical and chemical parameters of a substance, along with estimates of release patterns to the stream. In the second phase, tests are performed that evaluate the acute toxicity, or ability to kill outright, of substances likely to reach the river in substantial quantities. The third phase evaluates the chronic toxicity of the substances, or the probability that the substances will cause gradual, progressive damages to organisms. In the final phase, an estimate is made of the biological risk associated with a substance.[34]

In practice, there are numerous uncertainties associated with ecological risk assessment and management, including both gaps in the available data and uncertainties in the validity of model assumptions.[35] Models used to predict the biological activity of chemicals on the basis of their physical and chemical properties are statistical, rather than theoretical, in part because the effect of any given chemical may be a function of the presence or absence of other chemical contaminants. Sediment size, organic content, temperature, oxygen concentration, and pH all influence chemical behavior and toxicity, and individual species and life stages within a species may respond differently to the presence of a single contaminant. So, if we seek to evaluate the economic and biological costs of no action versus cleanup of the Argo Tunnel, for example, we must consider not only existing conditions, but also changes in water and sediment movement and water quality projected to result from increasing urbanization around Idaho Springs. Keeping in mind the multiple components which influence a river ecosystem, predicting the probable biological effects of a series of contaminants entering a river becomes an extremely complex task. However, this task must be attempted if we are to begin to understand the effects of our activities on stream biota.

One of the issues that must be kept in mind with regard to any type of water quality monitoring or prediction is that natural systems have background levels of spatial and temporal variability. These background levels must be known in order to determine whether human-caused disturbances are significantly altering a river.[36] A human-caused event may occur as a "press disturbance" that involves sustained or chronic interference with a natural population. An example would be alteration of flow regime as a result of dam construction. A "pulse disturbance" is an acute, short-term occurrence such as an accidental discharge of chemicals into the river. A "catastrophic disturbance" involves major, often planned, destruction of habitat such as channelization associated with road construction. The time-course of any disturbance is partly a function of the life cycle and longevity of potentially affected organisms. A disturbance may affect the mean abundance or temporal variance of populations, or the spatial distribution. These effects cannot be adequately assessed without a monitoring program that is both spatially diverse and relatively long-lasting. A pulse disturbance, such as the Exxon Valdez oil spill, attracts public attention for a few days or at most weeks. Once the incident fades from public memory, and some signs of recolonization occur, there is an implicit assumption that biological recovery has occurred. However, changes in the biotic community may become apparent over months or years. These changes cannot be documented without the type of sustained monitoring program that seems to test the patience of both the public and funding agencies. Unless we recognize that rivers are diverse in form and function, and that various types of disturbances may produce subtle, but important, physical, chemical, and biological alterations, we will not be able to preserve the Front Range rivers as anything but virtual rivers.

River Form Versus River Function

At the beginning of this book a question was posed: Is it feasible to save a fish without saving the ecosystem in which that fish evolved? Such a question is part of a larger issue, that of form versus function in rivers. Most people have at least a general expectation of the forms that a river may assume. Expectations for a "stable" or "healthy" river include vegetated banks that are not obviously eroding; a channel that is generally stationary in both its lateral and vertical dimensions at periods of years to decades; clear water; and a resident fish population. These qualities relate primarily to obvious, visual characteristics that affect people living along or using the river. However, such expectations provide only a simplistic assessment of a channel, and may be very misleading. The Front Range rivers provide a pertinent illustration. Even Clear Creek,

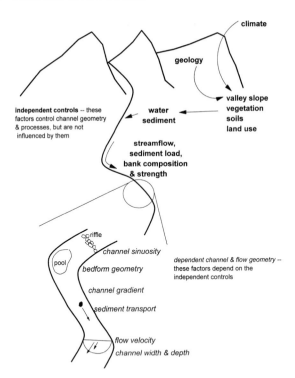

Schematic of interrelationships among physical processes in a river system.

climate

geology

independent controls -- these factors control channel geometry & processes, but are not influenced by them

water
sediment

valley slope
vegetation
soils
land use

streamflow,
sediment load,
bank composition
& strength

riffle

channel sinuosity

pool

bedform geometry

dependent channel & flow geometry --
these factors depend on the
independent controls

channel gradient

sediment transport

flow velocity
channel width & depth

the river most thoroughly disrupted by mining, and subsequently polluted by mining wastes, meets these criteria. But a closer examination reveals that Clear Creek has lost much of its function as an ecosystem. The streamside vegetation of the creek has decreased in density and diversity. The channel is partly stabilized by the bedrock valley walls and coarse bed-sediment, but also partly by highway and urban construction. The water of the creek is very clear, yet in some areas very toxic. And the trout in the creek are non-native species that are regularly stocked.

The form of a river may be defined as its physical characteristics, and these can be understood in terms of numerous levels of controlling variables. At the largest scales of space and time, geology and climate interact to determine the distributions of water and sediment reaching a river. Each aspect of river form reflects the distributions of water and sediment, from width/depth ratio of the stream channel through channel-bed gradient, to suspended sediment transport. At the most basic level, a river is a conduit for the downslope movement of water and sediment. But considered in detail, channel form resolves into a complex of delicately adjusted components.

Similarly, river function may be considered at varying levels of detail. At its simplest, the function of a river is to transport water and sediment. But living organisms, in all their amazing variety and abundance, have effectively colonized even the most minute and specialized niches within stream channels, so that it is appropriate to regard river function as including an ecosystem component. Just as a river impoverished in form may consist of a fairly uniform trench, so too may a river impoverished in function support a depauperate aquatic flora and fauna.[37] Those lacking specialized knowledge of rivers can be misled only too easily into assuming that a river is functioning well because its general form appears as expected.

These considerations are central to the third critical issue for the mountain rivers of the upper South Platte River basin in the twenty-first century: channel restoration. As land-use changes have resulted in changes to the water and sediment entering stream channels, these channels may become unsightly, pose a hazard to human life and property because of excessive scouring or sediment filling, or no longer provide some desired function, such as fishing. In this situation, some private landowners, municipalities, or conservation organizations have attempted to restore or stabilize portions of rivers, or even to relocate rivers. As noted in a U.S. Department of Transportation manual on this subject, "The ideal in habitat restoration would be to establish a stable riffle-pool sequence in the relocated channel."[38]

An example of successful channel stabilization comes from Badger Creek, a tributary of the Arkansas River in Colorado. This river was unstable as a result of increased water and sediment production associated with overgrazing, lumbering, roads, and mining.[39] River restoration was based on the use of hydraulic equations to design a stable channel with dimensions similar to those in less-disturbed reaches. Bank vegetation and metal cages filled with rock were used to reduce lateral channel migration, and a floodplain was incorporated to dissipate flow energy during large discharges. The initial channel was designed to be shallower and wider than the final desired channel in order to allow self-adjustment over a range of flow conditions.

Successful channel relocation or stabilization may be the exception rather than the norm. Often, check dams, deflectors, or rip-rap are placed without full knowledge or consideration of the hydraulic forces acting during large flows, resulting in the "blow-out" of these structures as described earlier for the Tenmile Creek project.[40] However, with increasing experience these efforts are likely to become more successful. A vital point to remember is that the river form has changed because controlling factors such as water and sediment discharge have changed. Any attempt to restore the river exactly to its predisturbance form is thus unlikely to succeed. Some allowance must be made for alterations in river form as a result of alterations in controlling factors. Similarly, the use of simple hydraulic geometry equations and idealized channel planforms from technical literature is not based in an understanding of local geology and channel characteristics, and represents an unrealistic and unsustainable attempt to force a river to assume an ideal form. A second vital point is that the river will not be static. Functioning rivers constantly undergo erosion and deposition as they adjust to changing discharges, and the stream restoration design must allow for this. The recognition that the various components of channel form are not the product of a single formative discharge, but rather of a range of discharges and of the temporal sequence of flows, is crucial in attempting to manage or restore rivers.

The approach of the City of Boulder to managing Boulder Creek provides an example of means by which urban alterations of stream channels may be minimized, and stream restoration efforts may be designed to include continuing channel adjustment. Boulder has purchased or designated open-space lands along the riparian corridor that serve as a buffer for the stream channel, as well as a means of preserving recreational sites. Along this corridor, city planners have replanted native riparian vegetation, enhanced fish habitat, and stocked trout. They have also designed foot/bicycle trails and bridges that direct traffic away from ecologically sensitive areas, and they have ensured instream flow protection.[41] Although channel movement within this riparian corridor might result in some loss of structures or landscaping, the costs of these damages would be much lower than the costs of completely stabilizing the channel and maintaining the stabilization structures, or the costs if residential or commercial structures built immediately beside the channel were damaged. More communities in the Colorado Front Range are now moving toward this model of riparian corridor protection, allowing rivers some measure of natural form and function, but the degree to which invasive engineering and structural control should enter into channel restoration remains controversial. As channel restoration increasingly becomes a business carried on by private consulting firms, there is a danger that restoration attempts may focus on imposing a cos-

The riparian corridor along Boulder Creek in the City of Boulder. The creek is at left. The city has designated a broad corridor of recreational open space and riparian vegetation along the creek.

In contrast to Boulder Creek, Clear Creek where it flows through Georgetown is narrowly constrained by riprap, buildings, and roads.

metic form on a river channel, rather than facilitating the re-creation of a functioning river ecosystem.

The issue of form versus function resolves into the question of what we expect from our rivers. Do we want to emphasize form at the grossest level, preserving rivers in which the flow regime is almost completely unrelated to the basin hydrology because of water transfers, for example? Or do we want function to assume equal importance, such that the rivers will be diverse, self-maintaining ecosystems? Unlike architecture, where form is said to follow function, the function of a river channel will follow directly from the form of that channel:[42] lacking woody debris and backwater sloughs, the macroinvertebrate and fish communities will be less diverse because of reduced habitat diversity; without a snowmelt-driven spring flood followed by low flows, a spatial zonation and increased diversity of streamside vegetation will not develop. The existence of endangered species, such as the native greenback cutthroat trout, is an indicator of problems in the function of an ecosystem. Because individual species have adapted through evolutionary processes to a specific landscape, rapid changes in that landscape may stress or eventually eliminate those species. The endangered trout of the Front Range rivers are a warning symbol, like the miner's canary, of the loss of river function.

The rivers of the Colorado Front Range have been impacted to varying degrees by numerous land-use activities during the past two centuries. The Poudre River described by Fremont in 1843, with lush streamside vegetation, numerous beaver dams, and abundant woody debris, is largely gone. As land-use activities have altered the forms of these rivers, their functions have been progressively reduced. The native greenback cutthroat trout is largely gone, too. We know almost nothing of the less conspicuous species, the macroinvertebrates, or reptiles, or algae, that may have gone with the trout. What is heartening is that these rivers do continue to function at some level. They still provide habitat for hundreds of species, and they still provide both the basis of physical life, and a rich emotional sustenance, for humans.

Respect for Rivers

Part of the problem is that there is relatively little incentive to change our land-use practices. As I read through the historical and technical accounts that form the basis of this book, I realized that we, as a culture,

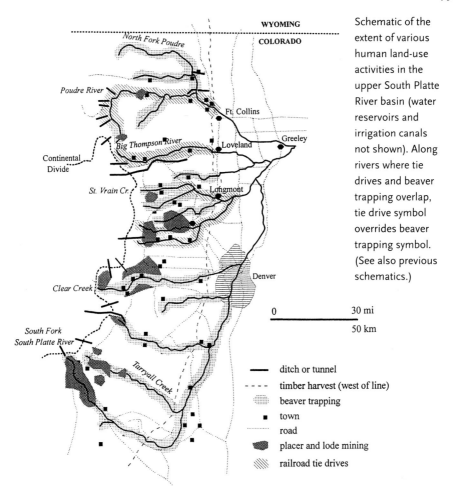

Schematic of the extent of various human land-use activities in the upper South Platte River basin (water reservoirs and irrigation canals not shown). Along rivers where tie drives and beaver trapping overlap, tie drive symbol overrides beaver trapping symbol. (See also previous schematics.)

WYOMING
COLORADO

North Fork Poudre

Poudre River

Ft. Collins

Greeley

Big Thompson River   Loveland

Continental
Divide

St. Vrain Cr.

Longmont

Denver

Clear Creek

South Fork
South Platte River

Tarryall Creek

0   30 mi

                    50 km

⎯⎯⎯ ditch or tunnel
- - - timber harvest (west of line)
▦ beaver trapping
■ town
······ road
▰ placer and lode mining
▨ railroad tie drives

have no respect for rivers. The industrious miners and timber-cutters of the 1860s were widely regarded as the stalwart pioneers of a superior civilization. Indeed, they were often hard-working, skilled, highly principled individuals.[43] But these people lived within a society that did not place a high value on natural ecosystems. Even those observers who decried the esthetic effects of placer mining seemed to regard the situation as a necessary evil, rather than an irresponsible and careless waste of natural resources. And today, if a canyon road needs to be straightened, or a new housing subdivision is planned for the foothills, there is too often only perfunctory consideration given to the effects on the Front Range rivers.

The situation is also complicated by our tradition of manipulating rivers for our own desires. Flood control through the use of dams and levees provides an obvious example, but an issue such as recreational fishing is also appropriate. The American Fish-Culturists' Association was established in 1870 to promote fish culture and widespread fish stocking. The association exemplified a utilitarian view of nature that perceived only benefits resulting from the spread of "good" sport fishes beyond their native range.[44] Today the United States has over a billion dollars of capital investment and numerous jobs involved with state, federal, and tribal hatcheries, and faces vocal constituencies of sport fishermen who often view native species as trash fish and who prefer to fish reservoirs or artificially maintained river reaches below dams. Natural resources managers responding to the demands of these fishermen may inadvertently participate in some startling inconsistencies among government agencies. Managers of the Coors brewery in Golden, Colorado, for example, are understandably upset when they are required to pay a fine for killing native white suckers from a beer spill, despite the fact that the Colorado Division of Wildlife pays to kill the same species with a fish toxin in order to improve the channel for sport fish (R. J. Behnke, pers comm, 1997).

Fishermen are not alone in viewing human-induced changes to rivers as beneficial. The changes in annual flow regime that changed the Great Plains portion of the South Platte River from a shallow, open stream to a river meandering between forested banks caused a loss of whooping crane habitat, but also created a habitat corridor into which white-tailed deer and eastern species of blue jays and orioles have moved. Gravel excavations from river floodplains in this region have formed ponds inhabited by introduced carp, and the carp feed herons, pelicans, cormorants, and eagles. In the mountains, river reaches below dams may have enhanced trout populations; trout biomass in the South Platte River below Cheeseman Dam is about ten times that of a typical population in natural waters (R. J. Behnke, pers comm, 1997). Human manipulation of rivers can thus benefit some species. It can be argued, however, that as we consistently manipulate natural environments to favor a relatively few "good" or resilient species, we are decreasing regional and global biodiversity. There is an amazing arrogance in this relentless simplification and homogenization of the natural world.

If a river is regarded as simply a conduit for water, with an occasional cottonwood for esthetic appeal, and a few hatchery trout for recreation, then there is no need to moderate our activities so as to preserve rivers in a natural state. This attitude seems to be prevalent in part because of widespread public ignorance of the effects of human activities on both the form and function of rivers. There is also little economic incentive to force the public to consider the efficiency of its water usage.

For example, when I moved to the city of Ft. Collins in 1989, most homes were not metered for water usage. The city has subsequently initiated a voluntary meter installation program. However, water costs remain very low. When I examined my residential water billing history for 1994, I discovered that it cost me only $2.40 more when my monthly water usage jumped 9.5 times (from 390 to 3,700 gallons, or 1,475–14,010 L). My water usage could increase by 4.8 times with no change in my bill. If my water usage increased more than a hundredfold, my bill only quadrupled. Under these conditions, decreased water consumption will obviously not be driven by economic necessity. This illustration is a minor one, but it relates to an argument that is increasingly applied to environmental issues. If most individuals and corporations act primarily for their short-term self-interest, allowing society and the environment to absorb the cost of their activities, then the most effective means of reducing environmental degradation is to restructure charges for natural resources so as to include the broader, truer range of costs.[45] This is not a straightforward task. What is the economic worth of a stonefly, or a native greenback cutthroat trout, relative to a pound of sugar beets or a green lawn? Such questions may seem ludicrous but they, or questions akin to them, must be considered. As population pressure and water demand continue to grow along the Front Range, we as a society will have to determine how highly we value functioning rivers. Because values are a human perception, they change through time. At present, values are changing toward greater appreciation of pristine natural environments and their recreational opportunities. Hopefully these changes can be carried through into economic incentives to preserve and restore the Front Range rivers.

As population pressures on Front Range water supplies increase, economics may play a more important role in development patterns. Water markets are not common in the United States, but have evolved in areas of the western United States such as the Colorado Front Range where water resources are scarce, water is privatized, and human populations are growing rapidly. In 1959, shortly after the Colorado-Big Thompson project began full delivery of water to the Northern Colorado Water Conservancy District, the District's board of directors decided that they would allow allotment contracts for water rights ownership to be rented on an annual basis or transferred permanently to other users as long as willing buyers and sellers existed. They made this decision in order to allow the highest degree of flexibility of this new supply to the greatest number of users.

This decision has prompted a number of water transfers from agriculture to municipalities in recent years. In terms of market economics, the highest value placed on water is for household consumption. Users are willing to pay the largest dollar amount for the first units of water made available to them, as these are of the greatest necessity to human survival. Typically, farmers are not willing or able to pay an equivalent sum for water, because water values are tied to crop mixes, crop prices, and efficiency rates of application. Thus, without administrative restrictions, water will always shift from agriculture to municipal use if driven by economic efficiency, as is presently occurring in the Colorado Front Range. In addition, water speculation is occurring as water is purchased with land that is not developed, the water being held and sold at a later date for a high market value.

At present, relative water prices are controlling rates and pattern of municipal growth in the northern Front Range urban corridor. If a developer wishes to build housing within Fort Collins, Greeley, or the Tri-districts (Fort Collins-Loveland Water District, East Larimer County Water District, and North Weld County Water District), the developer must either purchase water rights from a current holder and turn them over to one of these suppliers, or pay what is referred to as a "cash-in-lieu" rate to either one of the municipalities or to the water districts in exchange for treated water supply. Cash-in-lieu rates are based to large extent on current market prices for water from the Colorado-Big Thompson project. At present, the Tri-districts own only C-BT water rights, whereas Fort Collins and Greeley have access to Poudre River water. Development of property in the Tri-districts thus requires the purchase of C-BT water rights, which at present are five times what the cities of Fort Collins and Greeley have set as their cash-in-lieu rate. As a result, development in the Tri-districts is slower than that in the other areas.[46]

The primary intent behind this book has been to increase awareness of historic changes in the Front Range rivers resulting from human activities. This is not a story that will end when the last page of this book is read; rather, this is a review that may help to guide us in evaluating our future. I do not believe that it is feasible to save a fish without saving the ecosystem in which that fish evolved. We can invest large amounts of effort and money in saving a few individuals of an endangered species, but we cannot ultimately sustain the species over more than a few generations in the restricted environment of a zoo or an aquarium. If we are to preserve our natural inheritance, including the rivers that constitute such a vital portion of that inheritance, we must first understand the forms and functions of natural systems, and then regulate our own actions to preserve those forms and functions. In the Colorado Front Range, human population, and thus impacts on rivers, continues to grow. This situation is far from unique in the mountainous regions of the world, which have been identified as some of the most fragile and endangered natural environments as we enter the twenty-first century.[47] The issues of water quantity, water quality, and channel restoration that are critical to the future of rivers in the Colorado Front Range are also critical to the future of mountain rivers everywhere. We can go on as we have for much of the last 200 years, taking the most short-term and selfish view, grabbing as much as we can while it is still there, lulled into complacency by the apparently resilient form of these rivers. Or we can respect our rivers: we can recognize the destructive effects of our actions and consciously limit those actions. Only through self-restraint can we have rivers rather than virtual rivers. The integrity of a uniquely lovely and inspiring landscape rests on our choice.

Appendix

Conversion Factors for Metric to English Units

LENGTH

2.54 centimeters = 1 inch
1 meter = 3.28 feet
1 kilometer = 0.6 mile

WEIGHT

1 kilogram = 2.2 pounds
907.2 kilograms = 1 ton
31.3 grams = 1 troy ounce

AREA

1 hectare = 2.47 acres

VOLUME

1,233.5 cubic meters = 1 acre-foot
3.78 liters = 1 gallon

Notes

Chapter 1: Of Rivers and Virtual Rivers

1. J.C. Fremont. *Report of the exploring expedition to the Rocky Mountains in the year 1842, and to Oregon and North California in the years 1843-'44.* Washington: Gales and Seaton, Printers, 1845.

2. B. Taylor. *Colorado: a summer trip.* Niwot: University Press of Colorado, 1989 (1867).

3. J.F. Meline. *Two thousand miles on horseback: Santa Fe and back: a summer tour through Kansas, Nebraska, Colorado, and New Mexico in the year 1866.* Albuquerque: Horn and Wallace, 1966 (1868).

4. J.W. Berry. The climate of Colorado. In *Climates of the states,* v. 2. Port Washington, N.Y.: Water Information Center, Inc., 1974.

5. H.E. and M.A. Evans. *Cache la Poudre: the natural history of a Rocky Mountain river.* Niwot: University Press of Colorado, 1991; T.T. Veblen and D.C. Lorenz. *The Colorado Front Range: a century of ecological change.* Salt Lake City: University of Utah Press, 1991.

6. J. Jenik. "The diversity of mountain life." In B. Messerli and J.D. Ives, eds., *Mountains of the world: a global priority.* London: The Parthenon Publishing Group, London, 1997, pp. 199–231; J.D. Ives, B. Messerli and E. Spiess. "Mountains of the world—a global priority." In B. Messerli and J.D. Ives, eds., *Mountains of the world: a global priority.* London: The Parthenon Publishing Group, London, 1997, pp. 1–15.

7. M.E. Long. "Colorado's Front Range." *National Geographic Magazine* 190 (1996): 80–103.

8. D. Knighton. *Fluvial forms and processes.* London: Edward Arnold, 1984.

9. S.A. Schumm and R.S. Parker. 1973. "Implications of complex response of drainage systems for Quaternary alluvial stratigraphy." *Nature* 243 (1973): 99–100; G.K. Gilbert. "Hydraulic mining debris in the Sierra Nevada." *U.S. Geological Survey Professional Paper 105,* 1917; L.A. James. "Sustained storage and transport of hydraulic gold mining sediment in the Bear River, California." *Annals of the Association of American Geographers* 79 (1989): 570–592; L.A. James. "Incision and morphologic evolution of an alluvial channel recovering from hydraulic mining sediment." *Geological Society of America Bulletin* 103 (1991): 723–736; A.D. Knighton. "River adjustment to changes in sediment load: the effects of tin mining on the Ringarooma River, Tasmania, 1875–1984." *Earth Surface Processes and Landforms* 14 (1989): 333–359; A.D. Knighton. "Channel bed adjustment along mine-affected rivers of northeast Tasmania." *Geomorphology* 4 (1991): 205–219; M.M. Hilmes. *Changes in channel morphology associated with placer mining along the Middle Fork of the South Platte River, Fairplay, Colorado.* Ft. Collins: Unpublished M.S. thesis, Colorado State University, 1993.

10. Veblen and Lorenz. 1999.

11. J.A. Kennedy. *Cache la Poudre: Colorado's natural scenic river.* Ft. Collins: Unpublished Professional Paper, Colorado State University, 1967.

12. A. Brookes. "River channel change." In P. Calow and G.E. Petts, eds., *The rivers handbook: Hydrological and ecological principles.* Oxford: Blackwell Scientific Publications, 1994, v. 2, pp. 55–75.

13. M. Sandoz. *The beaver men.* Lincoln: University of Nebraska Press, 1964.

14. C. Bolgiano. *The Appalachian forest: a search for roots and renewal.* Mechanicsburg, Pa.: Stackpole Books, 1998.

15. J.D. Villamizar and J.M. Ramirez. *Caracteristicas de las explotaciones aluviales del Bajo Cauca y sus efectos ambeintales.* Medellin, Colombia: Universidad Nacional de Colombia, Tesis de Grado, 1988; B.P. Van Haveren. "Placer mining and sediment problems in interior Alaska." In *Proceedings of the Fifth Federal Interagency Sedimentation Conference,* 1991, v. 2, S.-S. Fan and Y.-H. Kuo, eds., pp. 10–69 to 10–74; Gilbert. 1917; James. 1989; James. 1991; Knighton. 1989; Knighton. 1991.

16. N. Smith. *A history of dams.* London: Peter Davies, 1971; E. Goldsmith and N. Hildyard. *The social and environmental effects of large dams.* San Francisco: Sierra Club Books, 1984; G.E. Petts. *Impounded rivers: perspectives for ecological management.* Chichester: John Wiley and Sons, 1984.

17. O. Hungr, G.C. Morgan, D.F. Van Dine, and D.R. Lister. "Debris flow defenses in British Columbia." In J.E. Costa and G.F. Wieczorek, eds., *Debris flows/avalanches: process, recognition, and mitigation.* Boulder, Colorado: Geological Society of America, Reviews in Engineering Geology, 1987, v. 7, pp. 201–236.

18. T. Mizuyama. "Structural and non-structural debris-flow countermeasures." In H.W. Shen, S.T. Su and F. Wen, eds., *Hydraulic engineering '93.* New York: ASCE, 1993, pp. 1914–1919: A. Armanini, F. Dellagiacoma, and L. Ferrari. "From the check dam to the development of functional check dams." In A. Armanini and G. DiSilvio, eds., *Fluvial hydraulics of mountain regions.* Braunschweig: Springer-Verlag, 1991, pp. 331–344; H.P. Willi. "Review of mountain river training procedures in Switzerland." In A. Armanini and G. Di Silvio, eds., *Fluvial hydraulics of mountain regions.* Braunschweig: Springer-Verlag, 1991, pp. 317–329; B. Wyzga. "Changes in the magnitude and transformation of flood waves subsequent to the channelization of the Raba River, Polish Carpathians." *Earth Surface Processes and Landforms* 21 (1996): 749–763.

19. C. Clark. "Deforestation and floods." *Environmental Conservation* 14 (1987): 67–69; L. Starkel. "Tectonic, anthropogenic and climatic factors in the history of the Vistula River valley downstream of Cracow." In G. Lang and C. Schluchter, eds., *Lake, mire, and river environments during the last 15,000 years.* Rotterdam: A.A. Balkema, 1988, pp. 161–170; K.G. Reinhart. "Effect of a commercial clearcutting operation in West Virginia on overland flow and storm runoff." *Journal of Forestry* 62 (1964): 162–172; W. F. Megahan and C.C. Bohn. "Progressive, long-term slope failure following road construction and logging on noncohesive, granitic soils of the Idaho Batholith." In *Headwaters Hydrology.* Bethesda, Maryland: American Water Resources Association, 1989, pp. 501–510; B.H. Heede. "Increased flows after timber harvest accelerate stream disequilibrium." In *Erosion control: a global perspective.* Proceedings of Conference XXII, International Erosion Control Association, 1991, pp. 449–454; J. D. Cheng. "Streamflow changes after clearcut logging of a pine-beetle infested watershed in southern British Columbia, Canada." *Water Resources Research* 25 (1989): 449–456; G.W. Brown and J.T. Krygier. "Clearcut logging and sediment production in the Oregon Coast Range." *Water Resources Research* 7 (1971): 1189–1198; B. Kasran. "Effect of logging on sediment yield in a hill Dipterocarp forest in Peninsular Malaysia." *Journal of Tropical Forest Science* 1 (1988): 56–66; A.R. Nik. "Water yield changes after forest conversion to agricultural landuse in Peninsular Malaysia." *Journal of Tropical Forest Science* 1 (1988): 67–84; Bolgiano. 1998.

20. G. Balamurugan. "Some characteristics of sediment transport in the Sungai Kelang Basin, Malaysia." *Journal of the Institution of Engineers, Malaysia* 48 (1991): 31–52; Y. Fukushima. "Estimating discharge and sediment yield from a forest road." In R.L. Beschta, T. Blinn, G.E. Grant, G.G. Ice, and F.J. Swanson, eds., *Erosion and sedimentation in the Pacific Rim*, Proceedings of the Corvallis Symposium, IAHS Publication no. 165, 1987, pp. 265–266; M.C. Larsen and J. E. Parks. "How wide is a road? the association of roads and mass-wasting in a forested montane environment." *Earth Surface Processes and Landforms* 22 (1997): 835–848; Bolgiano. 1998.

21. H.C. Pereira. *Policy and practice in the management of tropical watersheds*. Boulder, Colo.: Westview Press, 1989; J.S. Rawat and M.S. Rawat. "Accelerated erosion and denudation in the Nana Kosi watershed, Central Himalaya, India. part I: sediment load." *Mountain Research and Development* 14 (1994): 25–38; M. Carver and G. Nakarmi. "The effect of surface conditions on soil erosion and stream suspended sediments." In H. Schreier, P.B. Shah, and S. Brown, eds., *Challenges in mountain resource management in Nepal*. Kathmandu, Nepal: International Center for Integrated Mountain Development, 1995, pp. 155–162; T. Palmer. *Lifelines: the case for river conservation*. Washington, D.C.: Island Press, 1994; L.S. Hamilton and L.A. Bruijnzeel. "Mountain watersheds—integrating water, soils, gravity, vegetation, and people." In B. Messerli and J.D. Ives, eds., *Mountains of the world: a global priority*. London: The Parthenon Publishing Group, 1997, pp. 337–370; K. Klimek. "Man's impact on fluvial processes in the Polish Western Carpathians." *Geografiska Annaler* 69A (1987): 221–225; R.J. Chorley, S.A. Schumm and D.E. Sugden. *Geomorphology*. London: Methuen, 1984; M.S. Kearney and J.C. Stevenson. "Island land loss and marsh vertical accretion rate evidence for historical sea-level changes in Chesapeake Bay." *Journal of Coastal Research* 7 (1991): 403–415; Starkel. 1988.

22. R.J. Higgins, G. Pickup, and P.S. Cloke. "Estimating the transport and deposition of mining waste at Ok Tedi." In C.R. Thorne, J.C. Bathurst, and R.D. Hey, eds., *Sediment transport in gravel-bed rivers*. Chichester: John Wiley and Sons, 1987, pp. 949–976; W.R. Ismail and Z.A. Rahaman. "The impact of quarrying activity on suspended sediment concentration and sediment load of Sungai Relau, Pulau Pinang, Malaysia." *Malaysian Journal of Tropical Geography* 25 (1994): 45–57; Bolgiano. 1998.

23. F. Press and R. Siever. *Earth*, 4th ed. New York: W.H. Freeman and Co., 1986.

24. L. Schmidt. "Grow with the flow: growth and water in the South Platte basin." *Colorado Water* (newsletter of the Colorado Water Resources Research Institute), December 1997, pp. 9–10.

25. T.R. Eschner, R.F. Hadley, and K.D. Crowley. "Hydrologic and morphologic changes in channels of the Platte River Basin in Colorado, Wyoming, and Nebraska: a historical perspective." *U.S. Geological Survey Professional Paper 1277-A*, 1983.

26. W.C. Johnson. "Woodland expansion in the Platte River, Nebraska: patterns and causes." *Ecological Monographs* 64 (1994): 45–84.

27. C.T. Nadler and S.A. Schumm. "Metamorphosis of South Platte and Arkansas Rivers, eastern Colorado." *Physical Geography* (1981): 95–115; G.P. Williams. "The case of the shrinking channels—the North Platte and Platte Rivers in Nebraska." *U.S. Geological Survey Circular 781*, 1978; G.P. Williams. "Historical perspective of the Platte Rivers in Nebraska and Colorado." In W.D. Graul and S.J. Bissell, eds., *Lowland river and stream habitat in Colorado: a symposium*. Greeley: University of Northern Colorado, 1978, pp. 11–41.

28. S. Winckler. "The Platte pretzel." *Audubon* (1989): 86–112.

29. P.B. Bayley. "The flood pulse advantage and the restoration of river-floodplain systems." *Regulated Rivers: Research and Management* 6 (1991): 75–86; J.A. Gore and

F.D. Shields, Jr. "Can large rivers be restored?" *BioScience* 45 (1995): 142–152; R.E. Sparks. "Need for ecosystem management of large rivers and their floodplains." *BioScience* 45 (1995): 168–182; G.E. Petts, H. Moller, and A.L. Roux, eds. *Historical changes of large alluvial rivers: western Europe.* Chichester, Wiley and Sons, 1989.

30. M. Church. "Channel morphology and typology." In P. Calow and G.E. Petts, eds., *The rivers handbook: hydrological and ecological principles.* Oxford: Blackwell Scientific Publications, 1992, v. 1, pp. 126–143; E.A. Keller and W.N. Melhorn. "Rhythmic spacing and origin of pools and riffles." *Geological Society of America Bulletin* 89 (1978): 723–730; D.R. Montgomery and J.M. Buffington. "Channel reach morphology in mountain drainage basins." *Geological Society of America Bulletin* 109 (1997): 596–611.

31. R.D. Jarrett. "Flood elevation limits in the Rocky Mountains." In C.Y. Kuo, ed., *Engineering hydrology.* New York: American Society of Civil Engineers, 1993, pp. 180–185.

32. D. McComb. *Big Thompson: profile of a natural disaster.* Boulder, Colo.: Pruett Press, 1980.

33. N.W. Fry. *Cache la Poudre: "The River" as seen from 1889 by Norman Walter Fry.* Denver, Colo.: Private printing, 1954; R.R. Shroba, P.W. Schmidt, E.J. Crosby, W.R. Hansen, and J.M. Soule. "Geologic and geomorphic effects in the Big Thompson Canyon area, Larimer County. part B, storm and flood of July 31– August 1, 1976, in the Big Thompson River and Cache la Poudre River basins, Larimer and Weld Counties, Colorado." *U.S. Geological Survey Professional Paper* 1115, 1979, pp. 87–152.

34. S.W. Madsen. *Channel response associated with predicted water and sediment yield increases in northwest Montana.* Ft. Collins: Unpublished M.S. thesis, Colorado State University, 1994; Montgomery and Buffington. 1997.

35. E.E. Wohl. "Bedrock benches and boulder bars: floods in the Burdekin Gorge of Australia." *Geological Society of America Bulletin* 104 (1992): 770–778; J.E. O'Connor, R.H. Webb, and V.R. Baker. "Paleohydrology of pool-and-riffle pattern development: Boulder Creek, Utah." *Geological Society of America Bulletin* 97 (1986): 410–420; M.D. Harvey, R.A. Mussetter, and E.J. Wick. "A physical process-biological response model for spawning habitat formation for the endangered Colorado squawfish." *Rivers* 4 (1993): 114–131.

36. G.E. Petts and I. Maddock. "Flow allocation for in-river needs." In P. Calow and G.E. Petts, eds., *The rivers handbook: hydrological and ecological principles,* v. 2. Oxford: Blackwell Scientific Publications, 1994, pp. 289–307.

37. M.J. Winterbourn and C.R. Townsend. "Streams and rivers: one-way flow systems." In R.S.K. Barnes and K.H. Mann, eds., *Fundamentals of aquatic ecology,* 2nd ed. Oxford: Blackwell Scientific Publications, 1991, pp. 230–242.

38. R.F. Raleigh, L.D. Zuckerman, and P.C. Nelson. *Habitat suitability index models and instream flow suitability curves: brown trout.* U.S. Fish and Wildlife Service Biological Report 82(10.124), 1986; R.N.B. Campbell, D.M. Rimmer, and D. Scott. "The effect of reduced discharge on the distribution of trout." In A. Lillehammer and S.J. Saltveit, eds., *Regulated Rivers.* Oslo: Universitetsforlaget AS, 1984, pp. 407–416.

39. R.J. Behnke. "Native trout of western North America." *American Fisheries Society Monograph 6.* Bethesda, Md.: American Fisheries Society, 1992; J.T. Windell and S.Q. Foster. "The status of unexploited fish populations in the Green Lakes Valley, an alpine watershed, Colorado Front Range." In J.C. Halfpenny, ed., *Ecological studies in the Colorado alpine.* Boulder: Institute of Arctic and Alpine Research, University of Colorado, Occasional Paper 37, 1982, pp. 133–137; W.P. Dwyer and B.D. Rosenland. "Role of fish culture in the reestablishment of greenback cutthroat trout." *American*

Fisheries Society Symposium 4 (1988): 75–80; R.J. Stuber, B.D. Rosenlund, and J.R. Bennett. "Greenback cutthroat trout recovery program: management overview." *American Fisheries Society Symposium* 4 (1988): 71–74.

40. Behnke. 1992.

41. D.W. Reiser and R.G. White. "Effects of two sediment size-classes on survival of steelhead and chinook salmon eggs." *North American Journal of Fisheries Management* 8 (1988): 432–437.

42. R.B. Nehring. *Fish flow investigations.* Denver: Colorado Division of Wildlife, Stream Fisheries Investigations, Federal Aid in Sport Fish Restoration, Project F-51-R, Job 1, 1986.

43. U.S. Department of Agriculture Forest Service. *Cache la Poudre wild and scenic river: final environmental impact statement and study report.* Washington, D.C., 1980.

44. American Society of Civil Engineers Task Committee on Sediment Transport and Aquatic Habitats, Sedimentation Committee. "Sediment and aquatic habitat in river systems." *Journal of Hydraulic Engineering* 118 (1992): 669–687.

45. K.W. Cummins. "A review of stream ecology with special emphasis on organism-substrate relationships." In K.W. Cummins, C.A. Tryon, and R.T. Hartman, eds., *Organism-substrate relationships in streams.* Special Publication Number 4, Pymatuning Laboratory of Ecology, University of Pittsburgh, 1966, pp. 2–51.

46. J.A. Stanford and J.V. Ward. "The hyporheic habitat of river ecosystems." *Nature* 335 (1988): 64–66.

47. J.V. Ward. "A mountain river." In P. Calow and G.E. Petts, eds., *The rivers handbook: hydrological and ecological principles.* Oxford: Blackwell Scientific Publications, 1992, v. 1, pp. 493–510; J.V. Ward and B.C. Kondratieff. *An illustrated guide to the mountain stream insects of Colorado.* Niwot: University Press of Colorado, 1992.

48. A.M. Milner. "System recovery." In P. Calow and G.E. Petts, eds., *The rivers handbook: hydrological and ecological principles.* Oxford: Blackwell Scientific Publications, 1994, v. 2, p 76–97.

49. E.P. Odum. *Fundamentals of ecology,* 3rd ed. Philadelphia: W.B. Saunders Company, 1971.

50. R.J. Smith. "Managing grazing on riparian ecosystems to benefit wildlife." In O.B. Cope, ed., *Forum—Grazing and riparian/stream ecosystems.* Trout Unlimited, Inc., 1979, pp. 21–31; F.J. Wagstaff. "Economic issues of grazing and riparian area management." *Transactions 51st North American Wildlife and Natural Resources Conference* 51 (1986): 272–279; S.P. Cross. "Responses of small mammals to forest riparian perturbations." In *Riparian ecosystems and their management: reconciling conflicting uses.* First North American Riparian Conference, U.S.D.A. Forest Service General Technical Report RM-120, 1985, pp. 269–275; F.L. Knopf. "Significance of riparian vegetation to breeding birds across an altitudinal cline." In *Riparian ecosystems and their management: reconciling conflicting uses.* First North American Riparian Conference, U.S.D.A. Forest Service General Technical Report RM-120, 1985, pp. 105–111; K.W. Cummins and G.L. Spengler. "Stream ecosystems." *Water Spectrum* 10 (1978): 1–9; S.A. Schumm and D.F. Meyer. "Morphology of alluvial rivers of the Great Plains." In *Riparian and wetland habitats of the Great Plains.* Proceedings of the 31st Annual Meeting, Great Plains Agricultural Council Publication No. 91, 1979, pp. 9–15; B. Heede. "Interactions between streamside vegetation and stream dynamics." In *Riparian ecosystems and their management: reconciling conflicting uses.* First North American Riparian Conference, U.S.D.A. Forest Service General Technical Report RM-120, 1985, pp. 54–58.

51. R.L. Vannote, G.W. Minshall, K.W. Cummins, J.R. Sedell, and C.E. Cushing. "The river continuum concept." *Canadian Journal of Fisheries and Aquatic Sciences* 37 (1980): 130–137; J.V. Ward and J.A. Stanford. "The regulated stream as a testing ground for ecological theory." In A. Lillehammer and S.J. Saltveit, eds., *Regulated Rivers*. Oslo: Universitetsforlaget AS, 1984, pp. 23–38.

52. G.W. Minshall. "Autotrophy in stream ecosystems." *BioScience* 28 (1978): 767–771; G.W. Minshall, R.C. Petersen, T.L. Bott, C.E. Cushing, K.W. Cummins, R.L. Vannote, and J.R. Sedell. "Stream ecosystem dynamics of the Salmon River, Idaho: an 8th-order stream." *Journal North American Benthological Society* 11 (1992): 111–137; J.V. Ward. "Altitudinal zonation in a Rocky Mountain stream." *Arch. Hydrobiol./Suppl.* 74 2 (1986): 133–199; Ward and Kondratieff. 1992.

53. D.R. Montgomery. "Process domains and the river continuum." *Journal of the American Water Resources Association* 35 (1999): 397–410.

54. K.R. Fladmark. "Getting one's Berings." *Natural History* 95 (1986): 8–19; R. Gore. "The most ancient Americans." *National Geographic* 192 (1997): 92–99; J.M. Adovasio and R.C. Carlisle. "Pennsylvania pioneers." *Natural History* 95 (1986): 20–27; W.N. Irving. "New dates from old bones." *Natural History* 96 (1987): 8–13; T.D. Dillehay. "By the banks of the Chinchihuapi." *Natural History* 96 (1987): 8–12; N. Guidon. "Cliff notes." *Natural History* 96 (1987): 6–12; D. Stanford. "The Ginsberg experiment." *Natural History* 96 (1987): 10–14.

55. E.S. Cassells. *The archaeology of Colorado.* Boulder, Colo.: Johnston Books, 1983; V.T. Holliday. "Geoarchaeology and late Quaternary geomorphology of the Middle South Platte River, northeastern Colorado." *Geoarchaeology: An International Journal* 2 (1987): 317–329; W.M. Husted. "Prehistoric occupation of the alpine zone in the Rocky Mountains." In J.D. Ives and R.G. Barry, eds., *Arctic and alpine environments.* London: Methuen and Company, 1974, pp. 857–872; W.M. Husted. "Early occupation of the Colorado Front Range." *American Antiquity* 30 (1965): 494–498; J.B. Benedict. "Along the Great Divide: Paleoindian archaeology of the high Colorado Front Range." In D.J. Stanford and J.S. Day, eds., *Ice Age hunters of the Rockies.* Niwot: Denver Museum of Natural History and University Press of Colorado, 1992, pp. 343–359; J.B. Benedict. "Chronology of cirque glaciation, Colorado Front Range." *Quaternary Research* 3 (1973): 584–599; J.L. Eighmy. *Colorado plains prehistoric context for management of prehistoric resources of the Colorado Plains.* Denver: State Historical Society of Colorado, 1984; M.P. Grant. "A fluted projectile point from lower Poudre Canyon, Larimer County, Colorado." *Southwestern Lore* 54 (1988): 4–7; John Slay, U.S. Forest Service, Arapaho-Roosevelt National Forest archeologist, 10 January 1995, personal communication; "Mining the Rockies in 6000 B.C." *Discover* 15 (1994): 18.

56. D.R. Muhs. "Age and paleoclimatic significance of Holocene sand dunes in northeastern Colorado." *Annals of the Association of American Geographers* 75 (1985): 566–582; J.B. Benedict. "Prehistoric man and climate: the view from timberline." In R.P. Suggate and M.M. Creswell, eds., *Quaternary Studies.* Wellington: The Royal Society of New Zealand, 1975, pp. 67–74; J.B. Benedict and B.L. Olson. *The Mount Albion complex.* Ward, Colo.: Research Report No. 1, Center for Mountain Archeology, 1978; T.T. Veblen and D.C. Lorenz. *The Colorado Front Range: a century of ecological change.* Salt Lake City: University of Utah Press, 1991.

57. C.S. Marsh. *The Utes of Colorado: people of the Shining Mountains.* Boulder, Colo.: Pruett Publishing Company, 1982.

58. Cassells. 1983.

Chapter 2: The Beaver Men, 1811–1859

1. N.F. Payne. "Population dynamics of beaver in North America." *Acta Zoologica Fennica* 172 (1984): 263–266; A.W. Allen. *Habitat suitability index models: beaver*. Ft. Collins, Colo.: U.S. Fish and Wildlife Service, 1983; N.V. DeByle. "Wildlife." In N.V. DeByle and R.P. Winokur, eds., *Aspen: ecology and management in the western United States*. Ft. Collins, Colo.: U.S. Fish and Wildlife Service, Rocky Mountain Forest and Range Experiment Station, 1985, p. 135–152; D. Muchmore. "Beaver, if you will." *Wyoming Wildlife* 39 (1975): 16–21, 34.

2. E. James. *Account of an expedition from Pittsburgh to the Rocky Mountains*. Philadelphia: H.C. Carey and I. Lea, 1823.

3. J.C. Fremont. *Report of the exploring expedition to the Rocky Mountains in the year 1842, and to Oregon and North California in the years 1843-'44*. Washington, D.C.: Gales and Seaton, Printers, 1845, p. 282.

4. R.J. Naiman, J.M. Melillo, and J.E. Hobbie. "Ecosystem alteration of boreal forest streams by beaver (Castor canadensis)." *Ecology* 67 (1986): 1254–1269; R.J. Naiman, C.A. Johnston, and J.C. Kelley. "Alteration of North American streams by beaver." *BioScience* 38 (1988): 753–762.

5. M. Sandoz. *The beaver men: spearheads of empire*. Lincoln: University of Nebraska Press, 1964; W.H. Goetzmann. *Exploration and empire: the explorer and the scientist in the winning of the American West*. New York: W.W. Norton and Company, 1966; Naiman, Johnston, and Kelley. 1988.

6. T.J. Noel, P.F. Mahoney, and R.E. Stevens. *Historical atlas of Colorado*. Norman: University of Oklahoma Press, 1994; B. De Voto. *Across the wide Missouri*. Boston: Houghton Mifflin Co., 1947; Goetzmann. 1966.

7. P.C. Phillips. *The fur trade*, v. 2. Norman: University of Oklahoma Press, 1961; Goetzmann. 1966.

8. D.J. Wishart. *The fur trade of the American West 1807–1840: A geographical synthesis*. Lincoln: University of Nebraska Press, 1979; G.L. Peterson. *Four forts of the South Platte*. Fort Myer, Va.: Council on America's past, 1982; A. Watrous. *History of Larimer County, Colorado, 1911*. No location: The Old Army Press, 1972.

9. I.L. Bird. *A lady's life in the Rocky Mountains*. Norman: University of Oklahoma Press, 1960 (1878).

10. J.H. Tice. *Over the plains, on the mountains; or, Kansas, Colorado, and the Rocky Mountains, agriculturally, mineralogically, and aesthetically described*. St. Louis: Industrial Age Printing Company, 1872.

11. J.E. Grasse and E.F. Putnam. *Beaver management and ecology in Wyoming*. Cheyenne, Wyo.: Federal Aid in Wildlife Restoration Project, Bulletin No. 6, Wyoming Game and Fish Commission, 1955.; L.L. Rue. *The world of the beaver*. Philadelphia: Living World Books, J.B. Lippincott Co., 1964.

12. Naiman, Melillo, and Hobbie. 1986.

13. H.M. Dunning. *Over hill and vale: in the evening shadows of Colorado's Long's Peak*, v. 1. Boulder, Colo.: Johnson Publishing Company, 1956.

14. G. Hartman. "Habitat selection by European beaver (*Castor fiber*) colonizing a boreal landscape." *Journal of Zoology, London* 240 (1996): 317–325; A.M. Gurnell. "The hydrogeomorphological effects of beaver dam-building activity." *Progress in Physical Geography* 22 (1998): 167–189; D.R. Butler and G.P. Malanson. "Sedimentation rates and patterns in beaver ponds in a mountain environment." *Geomorphology* 13 (1995): 255–269; C.T. Driscoll, B.J. Wyskowski, C.C. Cosentini, and M.E. Smith. "Processes regulating temporal and longitudinal variations in the chemistry of a low-order wood-

land stream in the Adirondack region of New York." *Biogeochemistry* 3 (1987): 225–241.; K.L. Leidholt, W. McComb, and D.E. Hibbs. "The effect of beaver on stream and stream-side characteristics and coho populations in western Oregon." *Northwest Science* 63 (1989): 71; P.E. Busher, W. Randall, and J. Jenkins. "Population density, colony composition, and local movements in two Sierra Nevadan beaver populations." *Journal of Mammalogy* 64 (1983): 314; Naiman, Melillo, and Hobbie. 1986.

15. R. Olson and W.A. Hubert. *Beaver: water resources and riparian habitat manager.* Laramie: University of Wyoming, 1994.

16. D.R. Butler. *Zoogeomorphology: animals as geomorphic agents.* New York: Cambridge University Press, 1995.

17. H.F. Clifford, G.M. Wiley, and R.J. Casey. "Macroinvertebrates of a beaver-altered boreal stream of Alberta, Canada, with special reference to the fauna on the dams." *Canadian Journal of Zoology* 71 (1993): 1439–1447; A. Medwecka-Kornas and R. Hawro. "Vegetation on beaver dams in the Ojcow National Park (southern Poland)." *Phytocoenologia* 23 (1993): 611–618; P. Nummi. "Simulated effects of the beaver on vegetation, invertebrates and ducks." *Annales Zoologici Fennici* 26 (1989): 43–52.; J.D. Stock and I.J. Schlosser. "Short-term effects of a catastrophic beaver dam collapse on a stream fish community." *Environmental Biology of Fishes* 31 (1991): 123–129; Naiman, Johnston, and Kelley. 1988; Butler. 1995.

18. M. Parker. *Beaver, water quality, and riparian systems.* Laramie: Wyoming Water and Streamside Zone Conferences, Wyoming Water Research Center, University of Wyoming, 1986; S.D. Brayton. "The beaver and the stream." *Journal of Soil and Water Conservation* 39 (1984): 108–109; T.J. Maret, M. Parker, and T.E. Fannin. "The effect of beaver ponds on the nonpoint source water quality of a stream in southwestern Wyoming." *Water Research* 21 (1987): 263–268; M. Parker. "Relations among NaOH-extractable phosphorus, suspended solids, and ortho-phosphorus in streams of Wyoming." *Journal of Environmental Quality* 20 (1991): 271–278.

19. W.J. McNeil and W.H. Ahnell. *Success of pink salmon spawning relative to size of spawning bed materials.* U.S. Fish and Wildlife Service Special Science Report, Fish. 469, 1964; A.C. Cooper. *The effect of transported stream sediments on the survival of sockeye and pink salmon eggs and alevins.* Int. Pacific Salmon Fisheries Committee Bulletin 18, 1965.

20. F.H. Everest, R.L. Beschta, J.C. Scrivener, K.V. Koski, J.R. Sedell, and C.J. Cederholm. "Fine sediment and salmonid production: a paradox." In E.O. Salo and T.W. Cundy, eds., *Streamside management: forestry and fishery implications.* University of Washington, Institute of Forest Resources, Contribution No. 57, 1987, p. 98–142; D.D. Williams and J.H. Mundie. "Substrate size selection by stream invertebrates and the influence of sand." *Limnol. Oceanogr.* 23 (1978): 1020–1033.

21. K. Sullivan, T.E. Lisle, C.A. Dolloff, G.E. Grant, and L.M. Reid. "Stream channels: The link between forests and fishes." In E.O. Salo and T.W. Cundy, eds., *Streamside management: forestry and fishery implications.* University of Washington, Institute of Forest Resources, Contribution No. 57, 1987, p. 39–97; K.D. Fausch and T.G. Northcote. "Large woody debris and salmonid habitat in a small coastal British Columbia stream." *Canadian Journal of Fisheries and Aquatic Sciences* 49 (1992): 682–693; K.D. Fausch and R.G. Bramblett. "Disturbance and fish communities in intermittent tributaries of a western Great Plains River." *Copeia* 3 (1991): 659–674; T.E. Lisle. "Effects of aggradation and degradation on riffle-pool morphology in natural gravel channels, northwestern California." *Water Resources Research* 18 (1982): 1643–1651; M.L. Murphy and J.D. Hall. "Varied effects of clear-cut logging on predators and their habitat in small streams of the Cascade Mountains, Oregon." *Canadian Journal of Fisheries and*

Aquatic Science 38 (1981): 137–145; L.M. Reid and T. Dunne. "Sediment production from forest road surfaces." *Water Resources Research* 20 (1984): 1753–1761.

22. S.C. Williamson, J.M. Bartholow, and C.B. Stalnaker. "Conceptual model for quantifying pre-smolt production from flow-dependent physical habitat and water temperature." *Regulated Rivers: Research and Management* 8 (1993): 15–28; D.C. Erman, E.D. Andrews, and M. Yoder-Williams. "Effects of winter floods on fishes in the Sierra Nevada. *Canadian Journal of Fisheries and Aquatic Science* 45 (1988): 2195–2200.

23. Naiman, Melillo, and Hobbie. 1986.

Chapter 3: Civilization Comes to the Front Range, 1859–1990

1. C.W. Henderson. *Mining in Colorado: a history of discovery, development and production.* U.S. Geological Survey Professional Paper 138, 1926.

2. E.S. Bastin and J.M. Hill. *Economic geology of Gilpin County and adjacent parts of Clear Creek and Boulder Counties, Colorado.* U.S. Geological Survey Professional Paper 94, 1917.

3. P.W. Thrush (ed.). *A dictionary of mining, mineral, and related terms.* Washington, D.C.: U.S. Department of Interior, 1968.

4. M. Silva. *Placer gold recovery methods.* California Department of Conservation, Division of Mines and Geology, Special Publication 87, 1986.

5. Bastin and Hill. 1917.

6. W.F. Stone (ed.). *History of Colorado,* v. 1. Chicago: S.J. Clarke Publishing Company, 1918; B.H. Parker, Jr. *Gold placers of Colorado,* vol. 1. Golden: Colorado School of Mines, 1974.

7. Henderson. 1926.

8. A.R. Wallace. *Gold in the Central City Mining District, Colorado.* U.S. Geological Survey Bulletin 1857, chapter C, Gold-bearing polymetallic veins and replacement deposits—Part I, 1989, pp. C38-C47; Bastin and Hill. 1917; Silva. 1986.

9. J.A. Kennedy. *Cache la Poudre: Colorado's natural scenic river.* Ft. Collins: Unpublished professional paper, Colorado State University, 1967; K. Hess, Jr. *Rocky times in Rocky Mountain National Park: an unnatural history.* Niwot: University Press of Colorado, 1993; M.G. Montgomery. *A story of Gold Hill, Colorado: seventy-odd years in the heart of the Rockies.* Boulder, Colo.: The Book Lode, 1987; L.R. Haffen (ed.). *Overland routes to the gold fields, 1859, from contemporary diaries.* Glendale, Calif.: Arthur H. Clark Company, 1942; P. Smith. *A look at Boulder: from settlement to city.* Boulder, Colo.: Pruett Press, 1981; S. Pettem. *Red rocks to riches: gold mining in Boulder County, then and now.* Boulder, Colo.: Stonehenge Books, 1980; H.S. Cobb. *Prospecting our past: gold, silver and tungsten mills of Boulder County.* Boulder, Colo: The Book Lode, 1988; O.J. Hollister. 1973 (1867). *The mines of Colorado.* Arno Press, New York, 450 pp.; Henderson. 1926.

10. L.R. Haffen (ed.). *Colorado gold rush: contemporary letters and reports, 1858–1859.* Glendale, Calif.: Arthur H. Clark Company, 1941; S. Bowles. *Across the continent: a summer's journey to the Rocky Mountains, the Mormons, and the Pacific States, with Speaker Colfax.* Springfield, Mass.: Samuel Bowles and Company, 1865; E.D. Gardner and J.R. Guiteras. *Placer operations of Humphreys Gold Corporation, Clear Creek, Colorado.* U.S. Bureau of Mines Information Circular 6961, 1937; T. Cox. *Inside the mountains: a history of mining around Central City, Colorado.* Boulder, Colo.: Pruett Press, 1989; H.W.C. Prommel. "The Clear Creek placers and placer mining in Colorado." In *Guidebook: Rocky Mountain Association of Geologists Field Conference in Central Colo-*

rado, June 16–19, 1947. Rocky Mountain Association of Geologists, 1947, pp. 42–43; Henderson. 1926; Bastin and Hill. 1917.

11. D.A. Smith. *Colorado mining: a photographic history.* Albuquerque: University of New Mexico Press, 1977; *Colorado Bureau of Mines Scrapbooks,* 12 vols., v. 5, 1896–97, p. 84; Henderson. 1926; Bastin and Hill. 1917.

12. C.M. Clark. *A trip to Pike's Peak and notes by the way.* San Jose, Calif.: Talisman Press, 1958 (1861).

13. W.D. Bickham. *From Ohio to the Rocky Mountains: editorial correspondence of the Dayton (Ohio) Journal.* Dayton, Ohio: Journal Book and Job Printing House, 1879.

14. M.M. Hilmes. *Changes in channel morphology associated with placer mining along the Middle Fork of the South Platte River, Fairplay, Colorado.* Ft. Collins: Unpublished M.S. thesis, Colorado State University, 1993; Henderson. 1926.

15. M.O'C. Morris. *Rambles in the Rocky Mountains, with a visit to the gold fields of Colorado.* London: Smith, Elder, and Company, 1864.

16. J.F. Meline. *Two thousand miles on horseback: Santa Fe and back: a summer tour through Kansas, Nebraska, Colorado, and New Mexico in the year 1866.* Albuquerque: Horn and Wallace, 1966 (1868).

17. C. Harrington. *Summering in Colorado.* Denver: Richards and Company, 1874.

18. *History of Clear Creek and Boulder Valleys, Colorado.* Evansville, Ill: Unigraphic, Inc., 1971 (1880); J. Codman. *The round trip by way of Panama through California, Oregon, Nevada, Utah, Idaho, and Colorado, with notes on railroads, commerce, agriculture, mining, scenery, and people.* New York: G.P. Putnam's Sons, 1879; G.A. Crofutt. *Crofutt's grip-sack guide of Colorado,* v. 1. Omaha, Nebr.: Overland Publishing Company, 1881; A.D. Richardson. *Beyond the Mississippi: from the great river to the great ocean: life and adventure on the prairies, mountains, and Pacific coast, with more than two hundred illustrations, from photographs and original sketches, of the prairies, deserts, mountains, mines, cities, Indians, trappers, pioneers, and great natural curiosities of the new states and territories, 1857–1867.* Hartford, Conn.: American Publishing Company, 1867; B. Taylor. *Colorado: a summer trip.* Niwot: University Press of Colorado, 1989 (1867); W.R. Jones. *The dome of the continent: Colorado in 1872.* Olympic Valley, Calif.: Outbooks, 1977; Smith. 1977; Clark. 1958 (1861); Bickham. 1879; Meline. 1966 (1868); Harrington. 1874.

19. Smith. 1981.

20. Hilmes. 1993.

21. Gardner and Guiteras. 1937.

22. J.P. Wood. *Report on mine tailings pollution of Clear Creek, Clear Creek-Gilpin Counties, Colorado.* Denver: Report submitted to Davis and Wellbank, attorneys, 1935.

23. G. Greenwood. *New life in new lands: notes of travel.* New York: J.B. Ford and Company, 1873.

24. E.E. Van Nieuwenhuyse and J.D. LaPerriere. "Effects of placer gold mining on primary production in subarctic streams of Alaska." *Water Resources Bulletin* 22 (1986): 91–99; S.M. Wagener and J.D. LaPerriere. "Effects of placer mining on the invertebrate communities of interior Alaska streams." *Freshwater Invertebrate Biology* 4 (1985): 208–214; D.J. McLeay, I.K. Birtwell, G.F. Hartman, and G.L. Ennis. "Responses of Arctic Grayling (*Thymallus arcticus*) to acute and prolonged exposure to Yukon placer mining sediment." *Canadian Journal of Fisheries and Aquatic Sciences* 44 (1987): 658–673.

25. S.M. Haslam. *River pollution: an ecological perspective.* London: Belhaven Press, 1990.

26. H.H. Jackson. *Bits of travel at home*. New York: Little, Brown, and Company, 1898.

27. M.J. Winterbourn and C.R. Townsend. "Streams and rivers: one-way flow systems." In R.S.K. Barnes and K.H. Mann, eds., *Fundamentals of aquatic ecology*, 2nd ed. Oxford: Blackwell Scientific Publications, 1991, pp. 230–242.

28. S. Swales. "Environmental effects of river channel works used in land drainage improvement." *Journal of Environmental Management* 14 (1982): 103–126; Van Nieuwenhuyse and LaPerriere. 1985; McLeay, Birtwell, Hartman, and Ennis. 1987.

29. J. Lewin and M.G. Macklin. "Metal mining and floodplain sedimentation in Britain." In, V. Gardiner, ed., *International geomorphology 1986*, Part I. Chichester: Wiley, 1987, pp. 1009–1027.

30. G.K. Gilbert. *Hydraulic mining debris in the Sierra Nevada*. U.S. Geological Survey Professional Paper 105, 1917.

31. L.A. James. "Sustained storage and transport of hydraulic gold mining sediment in the Bear River, California." *Annals of the Association of American Geographers* 79 (1989): 570–592; A.D. Knighton. "River adjustment to changes in sediment load: the effects of tin mining on the Ringarooma River, Tasmania, 1875–1984." *Earth Surface Processes and Landforms* 14 (1989): 333–359; T.E. Lisle, J.E. Pizzuto, H. Ikeda, F. Iseya, and Y. Kodama. "Evolution of a sediment wave in an experimental channel." *Water Resources Research* 33 (1997): 1971–1981; A.D. Knighton. *Fluvial forms and processes*. London: Edward Arnold, 1984.

32. M.M. Hilmes and E.E. Wohl. "Changes in channel morphology associated with placer mining." *Physical Geography* 16 (1995): 223–242; Hilmes. 1993.

33. L.A. James. "Incision and morphologic evolution of an alluvial channel recovering from hydraulic mining sediment." *Geological Society of America Bulletin* 103 (1991): 723–736; L.A. James. "Channel changes wrought by gold mining: northern Sierra Nevada, California." In *Effects of human-induced changes on hydrologic systems*. Bethesda, Md.: American Water Resources Association, 1994. pp. 629–638; K.J. Fischer and M.D. Harvey. "Geomorphic response of lower Feather River to 19th century hydraulic mining operations." In *Inspiration: come to the headwaters. Proceedings, 15th Annual Conference of the Association of State Floodplain Managers, June 10–14, 1991*. Denver, Colo: 1991, pp. 128–132; A.D. Knighton. "Channel bed adjustment along mine-affected rivers of northeast Tasmania." *Geomorphology* 4 (1991): 205–219; B.P. Van Haveren. "Placer mining and sediment problems in interior Alaska." In S.S. Fan and Y.H. Kuo, eds., *Proceedings of the 5th Federal Interagency Sedimentation Conference*, v. 2, 1991, pp. 10–69 to 10–73; R.J. Higgins, G. Pickup, and P.S. Cloke. "Estimating the transport and deposition of mining waste at Ok Tedi." In C.R. Thorne, J.C. Bathurst, and R.D. Hey, eds., *Sediment transport in gravel-bed rivers*. Chichester: John Wiley and Sons, 1987, pp. 949–976; G. Parker, Y. Cui, J. Imran, and W.E. Dietrich. "Flooding in the lower Ok Tedi, Papua New Guinea due to the disposal of mine tailings and its amelioration." In *International seminar on recent trends of floods and their preventive measures, 20–21 June, 1996, Sapporo, Japan, Post-seminar proceedings*, Hokkaido River Disaster Prevention Research Center, 1997, pp. 21–46.

34. J.A. Stanford and F.R. Hauer. "Mitigating the impacts of stream and lake regulation in the Flathead River catchment, Montana, USA: an ecosystem perspective." *Aquatic Conservation: Marine and Freshwater Ecosystems* 2 (1992): 35–63.

35. K.M. Mackenthun and W.M. Ingram. "Pollution and the life in water." In K.W. Cummins, C.A. Tryon, and R.T. Hartman, eds., *Organism-substrate relationships in streams*. Special Publication No. 4, Pymatuning Laboratory of Ecology, University of Pittsburgh, 1966, pp. 136–145; J.V. Ward and B.C. Kondratieff. *An illustrated guide*

to the mountain stream insects of Colorado. Niwot: University Press of Colorado, 1992; Haslam. 1990.

36. J. Lewin, B.E. Davies, and P.J. Wolfenden. "Interactions between channel change and historic mining sediments." In K.J. Gregory, ed., River channel changes. Chichester: Wiley, 1977, pp. 353–367; A. Van Geen and Z. Chase. "Recent mine spill adds to contamination of southern Spain." EOS, Transactions of the American Geophysical Union 79 (1998): 449–455; M. Langedal. "The influence of a large anthropogenic sediment source on the fluvial geomorphology of the Knabeåna-Kvina rivers, Norway." Geomorphology 19 (1997): 117–132; W.L. Graf, S.L. Clark, M.T. Kammerer, T. Lehman, K. Randall, and R. Schröder. "Geomorphology of heavy metals in the sediments of Queen Creek." Catena 18 (1991): 567–582; J.C. Knox. "Historic valley floor sedimentation in the Upper Mississippi Valley." Annals of the Association of American Geographers 77 (1987): 224–244.

37. D.R. Ralston and A.G. Morilla. "Ground-water movement through an abandoned tailings pile." In R.F. Hadley and D.T. Snow, eds., Water resources problems related to mining. Minneapolis, Minn.: American Water Resources Association, 1974, pp. 174–183.

38. D.A. Wentz. Effects of mine drainage on the quality of streams in Colorado, 1971–72. Denver: Colorado Water Conservation Board, 1974; D.A. Wentz. "Stream quality in relation to mine drainage in Colorado." In R.F. Hadley and D.T. Snow, eds., Water resources problems related to mining. Minneapolis, Minn.: American Water Resources Association, 1974, pp. 158–173; T.R. Wildeman, D. Cain, and A.J. Ramirez. "The relation between water chemistry and mineral zonation in the Central City Mining District, Colorado." In R.F. Hadley and D.T. Snow, eds., Water resources problems related to mining. Minneapolis, Minn.: American Water Resources Association, 1974, pp. 219–229.

39. Haslam. 1990.

40. G.W. Bryan. "Some aspects of heavy metal tolerance in aquatic organisms." In A.P.M. Lockwood, ed., Effects of pollutants on aquatic organisms. Cambridge: Cambridge University Press, 1976, pp. 7–34.

41. G.M. Hughes. "Polluted fish respiratory physiology." In A.P.M. Lockwood, ed., Effects of pollutants on aquatic organisms. Cambridge: Cambridge University Press, 1976, pp. 163–183; Haslam. 1990; Bryan. 1976.

42. R.A. Roline and J.R. Boehmke. Heavy metals pollution of the upper Arkansas River, Colorado, and its effects on the distribution of the aquatic macrofauna. U.S. Department of Interior, Bureau of Reclamation, REC- ERC-81-15, 1981.

43. D.E. Rees. Indirect effects of heavy metals observed in macroinvertebrate availability, brown trout (Salmo trutta) diet composition, and bioaccumulation in the Arkansas River, Colorado. Ft. Collins: Unpublished M.S. thesis, Colorado State University, 1994; Roline and Boehmke. 1981. W.H. Clements, D.M. Carlisle, J.M. Lazorchak, and P.C. Johnson. "Heavy metals structure benthic communities in Colorado mountain streams." Ecological Applications 10 (2000): 626–638.

44. Environmental Protection Agency National Superfund Priorities List Sites: Colorado. Denver: U.S. E.P.A., Region VIII, 1994.

45. S.D. Machemer. Measurements and modeling of the chemical processes in a constructed wetland built to treat acid mine drainage. Golden, Colo.: Unpublished Ph.D. dissertation, Colorado School of Mines, 1992; C.M. Sellstone. Sequential extraction of Fe, Mn, Zn, and Cu from wetland substrate receiving acid mine drainage. Golden, Colo.: Unpublished M.S. thesis, Colorado School of Mines, 1990; J.C. Emerich, T.R. Wildeman, R.R. Cohen, and R.W. Klusman. "Constructed wetland treatment of acid mine

discharge at Idaho Springs, Colorado." In K.C. Stewart and R.C. Severson, eds., *Guidebook on the geology, history, and surface-water contamination and remediation in the area from Denver to Idaho Springs, Colorado.* U.S. Geological Survey Circular 1097, 1994, pp. 49–55.

46. R. Spitzer and G. Linden. "Geotechnical and environmental factors impacting redevelopment of a mining town, Central City, Colorado." In G.M. Norris and L.E. Meeker, eds., *Engineering geology and geotechnical engineering.* Reno, Nev.: 29th Symposium Proceedings, March 22–24, 1993, pp. 143–156.

47. J.M. Boyles, D. Cain, W. Alley, and R.W. Klusman. "Impact of Argo Tunnel acid mine drainage, Clear Creek County, Colorado." In R.F. Hadley and D.T. Snow, eds., *Water resources problems related to mining.* Minneapolis, Minn.: American Water Resources Association, 1974, pp. 41–53; W.H. Ficklin and K.S. Smith. "Influence of mine drainage on Clear Creek, Colorado." In K.C. Stewart and R.C. Severson, eds., *Guidebook on the geology, history, and surface-water contamination and remediation in the area from Denver to Idaho Springs, Colorado.* U.S. Geological Survey Circular 1097, 1994, pp. 43–48.

48. Wentz. 1974.

49. H.T. Stearns and L.W. Leroy. *Cause and effects of late Pleistocene and recent landslides near Georgetown, Colorado.* Boulder, Colo.: Geological Society of America Special Paper 82, 1965, p. 347; J.E. Costa and R.D. Jarrett. "Debris flows in small mountain stream channels of Colorado and their hydrologic implications." *Bulletin of the Association of Engineering Geologists* 18 (1981): 309–322.

50. Cox. 1989; Bickham. 1879; Meline. 1966 (1868); Crofutt. 1881; Jones. 1977; Greenwood. 1873.

51. Clark. 1958 (1861).

52. Haffen. 1942.

53. D.C. Kemp. *Silver, gold, and black iron: a story of the Grand Island Mining District of Boulder County, Colorado.* Denver: Sage Books, 1960; N.W. Fry. *Cache la Poudre: "The River" as seen from 1889 by Norman Walter Fry.* Private printing, 1954; E.A. Mills. *Early Estes Park.* 2nd ed. Big Mountain Press, 1963 (1911); A.H. Boardman. *Unpublished letters, 10 June to 10 October.* Denver Public Library, Western History Manuscript Collection, 1863; Montgomery. 1987; Richardson. 1867; Taylor. 1989 (1867); Jackson. 1898.

54. Haffen. 1941.

55. V.M Simmons. *The Upper Arkansas: a mountain river valley.* Boulder, Colo.: Pruett Publishing Company, 1990; J.H. Tice. *Over the plains, on the mountains; or, Kansas, Colorado, and the Rocky Mountains, agriculturally, mineralogically, and aesthetically described.* St. Louis, Mo.: Industrial Age Printing Company, 1872.

56. T.T. Veblen and D.C. Lorenz. *The Colorado Front Range: a century of ecological change.* Salt Lake City: University of Utah Press, 1991; K.M. Rowdabaugh. *The role of fire in the ponderosa pine-mixed conifer ecosystems.* Ft. Collins: Unpublished M.S. thesis, Colorado State University, 1978; R.D. Laven, P.N. Omi, J.G. Wyant, and A.S. Pinkerton. "Interpretation of fire scar data from a ponderosa pine ecosystem in the central Rocky Mountains, Colorado." In *Proceedings of the fire history workshop, Oct. 20–24, 1980, Tucson, Arizona.* Ft. Collins, Colo.: USDA Forest Service General Technical Report RM-81, Rocky Mountain Forest and Range Experiment Station, 1980, pp. 46–49; D. Goldblum. *Fire history of a ponderosa pine/Douglas-fir forest in the Colorado Front Range.* Boulder: Unpublished M.A. thesis, University of Colorado, 1990; Hess. 1993.

57. G.S. Henderson and D.L. Golding. "The effect of slash burning on the water repellency of forest soils at Vancouver, British Columbia." *Canadian Journal of For-*

est Research 13 (1983): 353–355; C.T. Dyrness. *Effect of soil wettability in the high Cascades of Oregon.* USDA Forest Service Research Paper PNW- 202, 1976; E.E. Wohl and P.A. Pearthree. "Debris flows as geomorphic agents in the Huachuca Mountains of southeastern Arizona." *Geomorphology* 4 (1991): 273–292; W.G. Wells. "The effects of fire on the generation of debris flows in southern California." In J.E. Costa and G.F. Wieczorek, eds., *Debris flows/avalanches: process, recognition, and mitigation.* Boulder, Colo.: Reviews in Engineering Geology, v. VII, Geological Society of America, 1987, pp. 105–114; J.L. Florsheim, E.A. Keller, and D.W. Best. "Fluvial sediment transport in response to moderate storm flows following chaparral wildfire, Ventura County, southern California." *Geological Society of America Bulletin* 103 (1991): 504–511.

58. *History of Clear Creek and Boulder Valleys.* 1971 (1880).

59. J.G. Rogers. *My Rocky Mountain valley.* Boulder, Colo.: Pruett Press, 1968.

60. B.H. Heede. "Response of a stream in disequilibrium to timber harvest." *Environmental Management* 15 (1991): 251–255; J.M. Bosch and J.D. Hewlett. "A review of catchment experiments to determine the effect of vegetation changes on water yield and evapotranspiration." *Journal of Hydrology* 55 (1982): 3–23; J.D. Cheng. "Streamflow changes after clearcut logging of a pine-beetle infested watershed in southern British Columbia, Canada." *Water Resources Research* 25 (1989): 449–456; C.A. Troendle and R.M. King. "The effect of partial and clearcutting on streamflow at Deadhorse Creek, Colorado." *Journal of Hydrology* 90 (1987): 145–157; R.G. Cline, H. Haupt, and G. Campbell. *Potential water yield response following clearcut harvesting on north and south slopes in northern Idaho.* U.S.D.A. Forest Service Research Paper INT-191, 1977; S. Berris and R.D. Harr. "Comparative snow accumulation and melt during rainfall in forested and clearcut plots in the western Cascades of Oregon." *Water Resources Research* 23 (1987): 135–142.

61. P.E. Black, E.C. Frank, R.H. Hawkins, R.C. Maloney, and J.R. Meiman. *Watershed analysis of the North Fork of the Cache la Poudre River, Larimer County, Colorado and Albany County, Wyoming.* Ft. Collins: Cooperative Watershed Management Unit, College of Forestry and Range Management, Colorado State University, 1959.

62. W.L. Graf. 1979. "Mining and channel response." *Annals of the Association of American Geographers,* v. 69, pp. 262–275.

63. M.G. Johnson and R.L. Beschta. "Logging, infiltration and surface erodibility in western Oregon." *Journal of Forestry* 78 (1980): 334–337; W.F. Megahan and W.J. Kidd. "Effects of logging and logging roads on erosion and sediment deposition from steep terrain." *Journal of Forestry* 70 (1972): 136–141; D.N. Swanston and F.J. Swanson. "Timber harvesting, mass erosion and steepland forest geomorphology in the Pacific Northwest." 1976; D.H. Gray and W.F. Megahan. *Forest vegetation removal and slope stability in the Idaho batholith.* U.S.D.A. Forest Service Research Paper INT-271, 1981; R. Ewing. "Postfire suspended sediment from Yellowstone National Park, Wyoming." *Journal of the American Water Resources Association* 32 (1996): 605–627.

64. W.S. Platts, R.J. Torquemada, M.L. McHenry, and C.K. Graham. "Changes in salmon spawning and rearing habitat from increased delivery of fine sediment to the South Fork Salmon River, Idaho." *Transactions American Fisheries Society* 118 (1989): 274–283; J.C. Scrivener. "Changes in composition of the streambed between 1973 and 1985, and the impacts on salmonids in Carnation Creek." In T.W. Chamberlin, ed., *Applying 15 years of Carnation Creek results.* Proceedings of the Workshop, 1988; T. Weaver and J. Frahley. "Fish habitat and fish populations." In *Final report—Flathead Basin Forest Practices—Water Quality and Fisheries Cooperative Program.* Kalispell, Mont.: Flathead Basin Commission, 1991, pp. 51–68.

65. J.K. Lyons and R.L. Beschta. "Land-use, floods and channel changes: Upper

Middle Fork Willamette River, Oregon (1936–1980)." *Water Resources Research* 19 (1983): 463–471; L.H. Powell. "Stream channel morphology changes since logging." In T.W. Chamberlin, ed., *Applying 15 years of Carnation Creek results.* Proceedings of the Workshop, 1987, pp. 16–25; D.W. Narver. *A survey of some possible effects of logging on two eastern Vancouver Island streams.* Fisheries Research Board of Canada Technical Report 323, 1972.

66. D.L. Hogan. *Channel morphology of unlogged, logged and debris torrented streams in the Queen Charlotte Islands.* British Columbia Ministry of Forests and Lands, Land Management Report No. 49, 1986; R.G. Roberts. *Stream channel morphology: major fluvial disturbances in logged watersheds on the Queen Charlotte Islands.* British Columbia Ministry of Forests, Land Management Report 48, 1987; Powell. 1987.

67. S.C. Ralph, G.C. Poole, L.L. Conquest, and R.J. Naiman. Stream channel morphology and woody debris in logged and unlogged basins of western Washington. *Canadian Journal of Fisheries and Aquatic Sciences* 51 (1994): 37–51; M.L. Murphy and J.D. Hall. Varied effects of clear-cut logging on predators and their habitat in small streams of the Cascade Mountains, Oregon. *Canadian Journal of Fisheries and Aquatic Sciences* 38 (1981): 137–145; S.W. Madsen. *Channel response associated with predicted water and sediment yield increases in northwest Montana.* Ft. Collins: Unpublished M.S. thesis, Colorado State University, 1994; Hogan. 1986.

68. A.D. Richmond. *Characteristics and function of large woody debris in mountain streams of northern Colorado.* Ft. Collins: Unpublished M.S. thesis, Colorado State University, 1994; L. Starkel. "Tectonic, anthropogenic and climatic factors in the history of the Vistula River valley downstream of Cracow." In G. Lang and C. Schluchter, eds., *Lake, mire and river environments during the last 15,000 years.* Rotterdam: A.A. Balkema, 1988, pp. 161–170; M.P. Mosley. "Erosion in the south-east Ruahine Range: its implications for downstream river control." New *Zealand Journal of Forestry* 23 (1978): 21–48; A.P. Brooks and G.J. Brierley. "Geomorphic responses of lower Bega River to catchment disturbance, 1851–1926." *Geomorphology* 18 (1997): 291–304; D. Vischer. "Impact of 18th and 19th century river training works: three case studies from Switzerland." In G.E. Petts, H. Moller, and A.L. Roux, eds., *Historical change of large alluvial rivers: western Europe.* Chichester: John Wiley and Sons, 1989, pp. 19–40; Z. Hugen and W. Jiquan. "The effect of forests on flood control — some comments on the flood disaster in the Huaihe River basin of Anhui Province in 1991." *Journal of Environmental Hydrology* 1 (1993): 38–43; G. Balamurugan. "Some characteristics of sediment transport in the Sungai Kelang Basin, Malaysia." *Journal of the Institution of Engineers, Malaysia* 48 (1991): 31–52; A.R. Nik. "Water yield changes after forest conversion to agricultural landuse in Peninsular Malaysia." *Journal of Tropical Forest Science* 1 (1988): 67–84; B. Kasran. "Effect of logging on sediment yield in a hill Dipterocarp forest in Peninsular Malaysia." *Journal of Tropical Forest Science* 1 (1988): 56–66.

69. Cox. 1989.

70. D.C. Renze. *A brief study of the lumber industry in Colorado, 1858–1948.* Unpublished term paper, 35 pp. Denver Public Library, Western History Collection, Manuscripts, 1949.

71. Richardson. 1867; Veblen and Lorenz. 1991.

72. Jones. 1977.

73. W.H. Wroten. *The railroad tie industry in the Central Rocky Mountain region: 1867–1900.* Ft. Collins: Unpublished Ph.D. dissertation, Colorado State University, 1956.

74. L.E. Leyendecker. *Georgetown: Colorado's silver queen, 1859–1876.* Ft. Collins, Colo: Centennial Publications, 1977; G. Morgan. *Three foot rails: a quick history of the*

Colorado Central Railroad. Private printing, 1974; J. Feldman. *An excursion up the South Platte Canyon: an informal history of the South Platte Canyon from Kessler to the town of South Platte.* Denver Board of Water Commissioners, Denver Public Library, Western History Collection, 1976; Henderson. 1926; Smith. 1981; Pettem. 1980; Roline and Boehmke. 1981.

75. Wroten. 1956.

76. Wroten. 1956.

77. R. Schmal and T. Wesche. "Historical implications of the railroad crosstie industry on current riparian and stream habitat management in the Central Rocky Mountains." In R.E. Gresswell, B.A. Barten, and J.L. Kershner, eds., *Practical approaches to riparian resource management: an educational workshop.* Billings, Mont.: U.S. Bureau of Land Management, 1989, p. 189.

78. Wroten. 1956.

79. Wroten. 1956.

80. Fry. 1954.

81. A. Watrous. *History of Larimer County, Colorado.* The Old Army Press, 1972 (1911); E.H. Ellis and C.S. Ellis. *The saga of Upper Clear Creek: a detailed history of an old mining area: its past and present.* Frederick, Colo.: Jende-Hagen Book Corporation, 1983.

82. H.M. Dunning. *Over hill and vale: in the evening shadows of Colorado's Long's Peak,* v.1. Boulder, Colo.: Johnson Publishing Company, 1956.

83. A. Ahlbrandt and K. Stieben (eds.). *The history of Larimer County, Colorado,* v.2. Dallas, Tex.: Curtis Media Corporation, 1987; H.M. Dunning. *Over hill and vale: in the evening shadows of Colorado's Long's Peak,* v.2. Dallas, Tex.: Curtis Media Corporation, 1956; *History of the Big Thompson Canyon.* Drake Home Demonstration Club; unpublished report in Denver Public Library Western History Collection, 1939.

84. L. Noble. "St. Vrain Valley pioneer recalls Lyons Tollhouse." Longmont, *Colorado Daily Times,* March 15–16, 1976, p. 5A.

85. Wroten. 1956.

86. M.K. Young, R.N. Schmal, and C.M. Sobczak. "Railroad tie drives and stream channel complexity: past impacts, current status, and future prospects." In *Proceedings of the Annual Meeting of the Society of American Foresters.* Bethesda, Md.: Publication 89-02, Society of American Foresters, 1990, pp. 126–130; M.K. Young, D. Haire, and M.A. Bozek. "The effect and extent of railroad tie drives in streams of southeastern Wyoming." *Western Journal of Applied Forestry* 9 (1994): 125–130; Schmal and Wesche. 1989.

87. F.G. Swanson, G.W. Lienkaemper, and J.R. Sedell. *History, physical effects, and management implications of large organic debris in western Oregon streams.* Portland, Ore.: U.S. Forest Service General Technical Report PNW-56, 1976; M.E. Harmon, J.F. Franklin, F.J. Swanson, P. Sollins, S.V. Gregory, J.D. Lattin, N.H. Anderson, S.P. Cline, N.G. Aumen, J.R. Sedell, G.W. Lienkaemper, K. Cromack, and K.W. Cummins. "Ecology of coarse woody debris in temperate ecosystems." *Advances in Ecological Research* 15 (1986): 133–302; E.A. Keller and T. Tally. "Effects of large organic debris on channel form and fluvial processes in the coastal redwood environment." In D.D. Rhodes and G.P. Williams, eds., *Adjustments of the fluvial system.* Dubuque, Iowa, Kendall/Hunt Publishing Company, 1979, pp. 169–197; R.A. Marston. "The geomorphic significance of log steps in forest streams." *Annals of the Association of American Geographers* 72 (1982): 99–108; E.A. Keller and F.J. Swanson. "Effects of large organic material on channel form and fluvial processes." Earth *Surface Processes* 4 (1979): 361–380; D.R. Bustard and D.W. Narver. "Aspects of the winter ecology of

juvenile coho salmon (*Oncorhynchus kisutch*) and steelhead trout (*Salmo gairdneri*)." *Journal of Fisheries Research Board of Canada* 32 (1975): 667–680; P.J. Tschaplinski and G.F. Hartman. "Winter distribution of juvenile coho salmon (*Oncorhynchus kisutch*) before and after logging in Carnation Creek, British Columbia, and some implications for overwinter survival." *Canadian Journal of Fisheries and Aquatic Sciences* 40 (1983): 452–461; M.L. Murphy, K.V. Koshi, J. Heifetz, S.W. Johnson, D. Kirchofer, and J.F. Thedinga. "Role of large organic debris as winter habitat for juvenile salmonids in Alaska streams." *Proceedings of the Western Association of Fish and Wildlife Agencies* 64 (1984): 251–262; M.D. Bryant. *Evolution of large, organic debris after timber harvest: Maybeso Creek, 1949 to 1978*. Portland, Ore.: U.S. Forest Service General Technical Report PNW-101, 1980; M.L. Murphy and J.D. Hall. "Varied effects of clear-cut logging on predators and their habitat in small streams of the Cascade Mountains, Oregon." *Canadian Journal of Fisheries and Aquatic Sciences* 38 (1981): 137–145; R.L. Beschta and W.S. Platts. "Morphological features of small streams: significance and function." *Water Resources Bulletin* 22 (1986): 369–379; E.G. Robison and R.L. Beschta. "Coarse woody debris and channel morphology interactions for undisturbed streams in southeast Alaska, USA." *Earth Surface Processes and Landforms* 15 (1990): 149–156; M.D. Bryant. "Changes 30 years after logging in large woody debris and its use by salmonids." In *U.S. Forest Service Technical Report RM-120*, Ft. Collins, Colo.: 1985, pp. 329–344; P.L. Angermeier and J.R. Karr. "Relationship between woody debris and fish habitat in a small warmwater stream." *Transactions of the American Fisheries Society* 113 (1984): 716–726; A.C. Benke, R.L. Henry, D.M. Gillespie, and R.J. Hunter. "Importance of snag habitat for animal production in southeastern streams." *Fisheries* 10 (1985): 8–13.

88. J.B. Minter. *Comparison of macroinvertebrate communities from three substrates in the lower South Platte River*. Ft. Collins: Unpublished MS thesis, Colorado State University, 1996.

89. A.D. Richmond and K.D. Fausch. "Characteristics and function of large woody debris in subalpine Rocky Mountain streams in northern Colorado." *Canadian Journal of Fisheries and Aquatic Sciences* 52 (1995): 1789–1802.

90. R.E. Bilby. "Removal of woody debris may affect stream channel stability." *Journal of Forestry* (1984): 609- 613; D.L. Hogan. "The influence of large organic debris on channel recovery in the Queen Charlotte Islands, British Columbia, Canada." In *Erosion and sedimentation in the Pacific Rim*, IAHS Publication no. 165, 1984, pp. 343–353; K.D. Fausch and T.G. Northcote. "Large woody debris and salmonid habitat in a small coastal British Columbia stream." *Canadian Journal of Fisheries and Aquatic Sciences* 49 (1992): 682–693; T.E. Lisle. "Effects of coarse woody debris and its removal on a channel affected by the 1980 eruption of Mount St. Helens, Washington." *Water Resources Research* 31 (1995): 1797–1808; D.R. Montgomery, J.M. Buffington, R.D. Smith, K.M. Schmidt, and G. Pess. "Pool spacing in forest channels." *Water Resources Research* 31 (1995): 1097–1105; K.J. Gregory, A.M. Gurnell, and C.T. Hill. "The permanence of debris dams related to river channel processes." *Hydrological Sciences Journal* 30 (1985): 371–381; P.A. Bisson, R.E. Bilby, M.D. Bryant, C.A. Dolloff, G.B. Grette, R.A. House, M.L. Murphy, K.V. Koski, and J.R. Sedell. "Large woody debris in forested streams in the Pacific Northwest: past, present, and future." In E.O. Salo and T.W. Cundy, eds., *Streamside management: forestry and fishery implications*. University of Washington, Institute of Forest Resources, Contribution No. 57, 1987, pp. 143–190; R.E. Bilby and J.W. Ward. "Changes in characteristics and function of woody debris with increasing size of streams in western Washington." *Transactions of the American Fisheries Society* 118 (1989): 368–378; G.P. Malanson and D.R. Butler. "Woody de-

bris, sediment and riparian vegetation of a subalpine river, Montana, USA." *Arctic and Alpine Research* 22 (1990): 183–194; F. Nakamura and F.J. Swanson. "Effects of coarse woody debris on morphology and sediment storage of a mountain stream system in western Oregon." *Earth Surface Processes and Landforms* 18 (1993): 43–61.

91. Wroten. 1956.

92. Watrous. 1972 (1911).

93. Meline. 1966 (1868).

94. S. Gerlek. *Water supplies of the Platte River basin.* Ft. Collins: Unpublished M.S. thesis, Colorado State University, 1977; J.E. Field. "Development of irrigation." In W.F. Stone, ed., *History of Colorado*, v.1. Chicago: S.J. Clarke Publishing Company, 1918, pp. 491–505; T.J. Noel, P.F. Mahoney, and R.E. Stevens. *Historical atlas of Colorado.* Norman: University of Oklahoma Press, 1994.

95. E.B. Swanson. *Red Feather Lakes: the first hundred years.* Private printing, 1971.

96. Feldman. 1976.

97. W.E. Pabor. *Colorado as an agricultural state: its farms, fields, and garden lands.* New York: Orange Judd Company, 1883.

98. H.E. Evans and M.A. Evans. *Cache la Poudre: the natural history of a Rocky Mountain river.* Niwot: University Press of Colorado, 1991; Kemp. 1960; Black, Frank, Hawkins, Maloney, and Meiman. 1959; History of the Big Thompson Canyon. 1939; Gerlek. 1977.

99. F. Fossett. *Colorado: its gold and silver mines, farms and stock ranges, and health and pleasure resorts: tourist's guide to the Rocky Mountains.* New York: C.G. Crawford, 1880.

100. National Park Service, Rocky Mountain Regional Office. *Resource assessment: Proposed Cache la Poudre River National Heritage Corridor.* National Park Service D-106, 1990; D. Tyler. *The last water hole in the West: the Colorado-Big Thompson project and the Northern Colorado Water Conservancy District.* Niwot: University Press of Colorado, 1992; J.S. Lochhead. *Transmountain water diversions in Colorado.* Occasional Papers Series, University of Colorado, Boulder, Natural Resources Law Center, 1987; Gerlek. 1977.

101. Tyler. 1992.

102. Tyler. 1992.

103. *Colorado-Big Thompson Project, technical record of design and construction. vol. 1: planning, legislation, and general description.* Denver: U.S. Bureau of Reclamation, 1957; M. Reisner and S. Bates. *Overtapped oasis: reform or revolution for western water.* Washington, D.C.: Island Press, 1990; Tyler. 1992.

104. A.H. Allen. "Pioneer life in Old Burlington." *Colorado Magazine* 14 (1937): 155–156. (From Wroten, 1956)

105. Black, Frank, Hawkins, Maloney, and Meiman. 1959.

106. R.D. Jarrett. "Flood elevation limits in the Rocky Mountains." In C.Y. Kuo, ed., *Engineering hydrology.* New York: American Society of Civil Engineers, 1993, pp. 180–185.

107. R.A. Schleusener, G.L. Smith, and M.C. Chen. "Effect of flow diversion for irrigation on peak rates of runoff from watersheds in and near Rocky Mountain foothills of Colorado." *International Association of Hydrologists Bulletin* 7 (1962): 53–61.

108. S.E. Ryan. *Effects of transbasin diversion on flow regime, bedload transport, and channel morphology in Colorado mountain streams.* Boulder: Unpublished Ph.D. dissertation, University of Colorado, 1994.

109. G.P. Williams and M.G. Wolman. "Effects of dams and reservoirs on surface-water hydrology—changes in rivers downstream from dams." *National Water Sum-*

mary 1985 — Water-Availability Issues, 1985, pp. 83–88; R.F. Hadley and W.W. Emmett. "Effects of dam construction on channel geometry and bed material in Bear Creek, Denver, Colorado." In G.D. Glysson, ed., *Proceedings of the advanced seminar on sedimentation, August 15–19, 1983, Denver, Colorado.* U.S. Geological Survey Circular 953, 1987, p. 35.

110. J.A. Stanford. *Ecological studies of Plecoptera in the Upper Flathead and Tobacco Rivers, Montana.* Salt Lake City: Unpublished Ph.D. dissertation, University of Utah, 1975; F.R. Hauer and J.A. Stanford. "Ecology and life histories of three net-spinning caddisfly species (Hydropsychidae: Hydropsyche) in the Flathead River, Montana." *Freshwater Invertebrate Biology* 1 (1982): 18–29; J.J. Fraley, B. Marotz, and J. DosSantos. *Fisheries mitigation plan for losses attributable to the construction and operation of Hungry Horse Dam.* Kalispell, Mont.: Montana Department of Fish, Wildlife, and Parks and the Confederated Salish and Kootenai Tribes, 1991; F.R. Hauer. *Ecological studies of Trichoptera in the Flathead River, Montana.* Denton, Tex.: Unpublished Ph.D. dissertation, North Texas State University, 1980; J.V. Ward and J.A. Stanford. "Thermal responses in the evolutionary ecology of aquatic insects." *Annual Review of Entomology* 27 (1982): 97–117; J.A. Gore. "Hydrological change." In P. Calow and G.E. Petts, eds., *The rivers handbook: Hydrological and ecological principles,* v. 2. Oxford: Blackwell Scientific Publications, 1994, pp. 33–54.

111. H.J. Zimmermann and J.V. Ward. "A survey of regulated streams in the Rocky Mountains of Colorado, USA." In A. Lillehammer and S.J. Saltveit, eds., *Regulated rivers.* Oslo: Universitetsforlaget AS, 1984, pp. 251–262.

112. American Society of Civil Engineers Task Committee on Sediment Transport and Aquatic Habitats, Sedimentation Committee. "Sediment and aquatic habitat in river systems." *Journal of Hydraulic Engineering* 118 (1992): 669–687; Ward and Kondratieff. 1992.

113. Stanford and Hauer. 1992.

114. S.A. Perry and W.B. Perry. "Effects of experimental flow regulation on invertebrate drift and stranding in the Flathead and Kootenai Rivers, Montana, USA." *Hydrobiologia* 134 (1986): 171–182; J.J. Fraley and J. Decker-Hess. "Effects of stream and lake regulation on reproductive success of kokanee salmon in the Flathead system, Montana." *Regulated Rivers: Research and Management* 1 (1982): 257–265; P.D. Armitage. "Environmental changes induced by stream regulation and their effect on lotic macroinvertebrate communities." In A. Lillehammer and S.J. Saltveit, eds., *Regulated rivers.* Oslo: Universitetsforlaget AS, 1984, pp. 139–165; Stanford and Hauer. 1992; Stanford. 1975.

115. R.R. Harris, R.J. Risser, and C.A. Fox. "A method for evaluating streamflow discharge-plant species occurrence patterns on headwater streams." In *Riparian ecosystems and their management: reconciling conflicting uses.* First North American Riparian Conference, U.S.D.A. Forest Service General Technical Report RM-120, 1985, pp. 87–90; K.S. Richards, F.M.R. Hughes, A.S. El-Hames, T. Harris, G. Pautou, J.-L. Peiry, and J. Girel. 1996. "Integrated field, laboratory and numerical investigations of hydrological influences on the establishment of riparian tree species." In *Floodplain processes,* M.G. Anderson, D.E. Walling, and P.D. Bates, eds. Chichester: Wiley and Sons, 1996, pp. 611–635; I.D. Jolly. "The effects of river management on the hydrology and hydroecology of arid and semi-arid floodplains." In *Floodplain processes,* M.G. Anderson, D.E. Walling, and P.D. Bates, eds. Chichester: Wiley and Sons, 1996, pp. 577–609.

116. J.C. Stromberg and D.T. Patten. "Instream flow requirements for riparian vegetation." In *Legal, institutional, financial and environmental aspects of water issues.* Am. Society of Civil Engineers Proceedings, July 1989, University of Delaware, 1989,

pp. 123–130; E.S. Menges and D.M. Waller. "Plant strategies in relation to elevation and light in floodplain herbs." *The American Naturalist* 122 (1983): 454–473; C.R. Hupp and W.R. Osterkamp. "Riparian vegetation and fluvial geomorphic processes." *Geomorphology* 14 (1996): 277–295.

117. L. Pearlstine, H. McKellar, and W. Kitchens. "Modelling the impacts of a river diversion on bottomland forest communities in the Santee River floodplain, South Carolina." *Ecological Modelling* 29 (1985): 283–302; W.C. Johnson. "Dams and riparian forests: case study from the upper Missouri River." In *Restoration, creation, and management of wetland and riparian ecosystems in the American West, Denver, Colorado,* 1988; C. Nilsson. "Remediating river margin vegetation along fragmented and regulated rivers in the North: what is possible?" *Regulated Rivers: Research and Management* 12 (1996): 415–431; J.C. Stromberg and D.T. Patten. "Riparian vegetation instream flow requirements: a case study from a diverted stream in the eastern Sierra Nevada, California, USA." *Environmental Management* 14 (1990): 185–194; D.M. Merritt. *The effects of mountain reservoir operations on the distributions and dispersal mechanisms of riparian plants, Colorado Front Range.* Ft. Collins: Unpublished Ph.D. dissertation, Colorado State University, 1999.

118. M.L. Scott, J.M. Friedman, and G.T. Auble. "Fluvial process and the establishment of bottomland trees." *Geomorphology* 14 (1996): 327–339.

119. L.D. Cline and J.V. Ward. "Biological and physicochemical changes downstream from construction of a subalpine reservoir, Colorado, USA." In A. Lillehammer and S.J. Saltveit, eds., *Regulated rivers.* Oslo: Universitetsforlaget AS, 1984, pp. 233–243.

120. L.J. Gray and J.V. Ward. *Effects of releases of sediment from reservoirs on stream biota.* Ft. Collins, Colo: Technical Completion Report B-226-COLO, Colorado Water Resources Research Institute, 1982.

121. E.E. Wohl and D.A. Cenderelli. "Sediment deposition and transport patterns following a reservoir sediment release." *Water Resources Research* 36 (2000): 319–333.

122. R.E. Zuellig, B.C. Kondratieff, and H.A. Rhodes. "Benthos recovery after an episodic sediment release into a Colorado Rocky Mountain river, USA." *Western Naturalist.* In press.

123. Stanford and Hauer. 1992.

124. D. Miller, Operations Coordinator, Northern Colorado Water Conservancy District. January 24, 1995, personal communication.

125. T.J. Sheets. "Agricultural pollutants." In F.E. Guthrie and J.J. Perry, eds., *Introduction to environmental toxicology.* New York: Elsevier, 1980, pp. 24–33.

126. Clark. 1958 (1861); Tice. 1872; Morgan. 1974; Ellis and Ellis. 1983.

127. Harrington. 1874.

128. Ellis and Ellis. 1983.

129. K.G. Reinhart. "Effect of a commercial clearcutting operation in West Virginia on overland flow and storm runoff." *Journal of Forestry* 62 (1964): 162–172; M.G. Johnson and R.L. Beschta. "Logging, infiltration and surface erodibility in western Oregon." *Journal of Forestry* 78 (1980): 334–337.

130. J.D. Balog. *Big Thompson River tributaries: geomorphic activity and its controlling factors during the 1976 flood.* Boulder: Unpublished M.A. thesis, University of Colorado, 1977.

131. G.W. Brown and J.T. Krygier. "Clearcut logging and sediment production in the Oregon Coast Range." *Water Resources Research* 7 (1971): 1189–1198; L.M. Reid and T. Dunne. "Sediment production from forest road surfaces." *Water Resources Research* 20 (1984): 1753–1761; B. Lorch. *Transport and aquatic impacts of highway traction sand*

and salt near Vail Pass, Colorado. Ft. Collins: Unpublished M.S. thesis, Colorado State University, 1998.

132. L.D. Cline, R.A. Short, and J.V. Ward. "The influence of highway construction on the macroinvertebrates and epilithic algae of a high mountain stream." *Hydrobiologia* 96 (1982): 149–159.

133. D.N. Swanston and F.J. Swanson. "Timber harvesting, mass erosion and steepland forest geomorphology in the Pacific Northwest." In D.R. Coates, ed., *Geomorphology and engineering.* Dowden, Hutchinson & Ross, 1976, pp. 199–221; M.C. Larsen and J.E. Parks. "How wide is a road? the association of roads and mass-wasting in a forested montane environment." *Earth Surface Processes and Landforms* 22 (1997): 835–848.

134. U.S. Department of Agriculture Forest Service. *Cache la Poudre wild and scenic river: final environmental impact statement and study report,* 1980; Black, Frank, Hawkins, Maloney, and Meiman. 1959.

135. F.E. Busby. "Riparian and stream ecosystems, livestock grazing, and multiple-use management." In O.B. Cope, ed., *Forum—grazing and riparian/stream ecosystems.* Trout Unlimited, Inc., 1979, pp. 1–6; R.J. Smith. "Managing grazing on riparian ecosystems to benefit wildlife." In O.B. Cope, ed., *Forum—grazing and riparian/stream ecosystems.* Trout Unlimited, Inc., 1979, pp. 21–31.

136. Harrington. 1874.

137. I.E. Wallin. *The professors' ranch.* Boulder, Colo: Johnson Publishing Company, 1964.

138. D.A. Wentz. *Environment of the middle segment, Cache la Poudre River, Colorado.* Denver: Colorado Division of Wildlife, 1974.

139. R. Van Buren. Fishery Biologist, Colorado Division of Wildlife, 18 January 1995, personal communication.

140. C.N. Spencer, B.R. McClelland, and J.A. Stanford. "Shrimp stocking, salmon collapse, and eagle displacement." *BioScience* 41 (1991): 14–21; D.W. Chess, F. Gibson, A.T. Scholz, and R.J. White. "The introduction of lahontan cutthroat trout into a previously fishless lake: feeding habits and effects upon the zooplankton and benthic community." *Journal of Freshwater Ecology* 8 (1993): 215–225.

141. B. Hodgson. "Buffalo: back home on the range." *National Geographic* 186 (1994): 64–89.

142. Hess. 1993.

143. Stone. 1918; Hess. 1993; Black, Frank, Hawkins, Maloney, and Meiman. 1959; Ellis and Ellis. 1983; Hodgson. 1994.

144. M. Estes. *The memoirs of Estes Park.* Ft. Collins: Friends of the Colorado State College Library, 1939; H.L. Gary, S.R. Johnson, and S.L. Ponce. "Cattle grazing impact on surface water quality in a Colorado Front Range stream." *Journal of Soil and Water Conservation* 38 (1983): 124–128; F.J. Wagstaff. "Economic issues of grazing and riparian area management." *Transactions, 51st North American Wildlife and Natural Resources Conference* 51 (1986): 272–279.

145. E.A. Dahlem. "The Mahogany Creek watershed—with and without grazing." In O.B. Cope, ed., *Forum—grazing and riparian/stream ecosystems.* Trout Unlimited, Inc., 1979, pp. 31–34; W.H. Blackburn, R.W. Knight, and M.K. Wood. *Impacts of grazing on watersheds: a state of knowledge.* College Station: The Texas Agricultural Experiment Station, The Texas Agriculture and Mining University, MP 1496, 1982; W.S. Platts. "Livestock grazing and riparian/stream ecosystems—an overview." In O.B. Cope, ed., *Forum—grazing and riparian/stream ecosystems.* Trout Unlimited, Inc., 1979, pp. 39–45; J.B. Kauffman and W.C. Krueger. "Livestock impacts on riparian ecosystems and

streamside management implications . . . a review." *Journal of Range Management* 37 (1984): 430–438; C. Clifton. "Effects of vegetation and land use on channel morphology." In R. Gresswell, B. Barton, and J. Kershner, eds., *Practical approaches to riparian resource management: an educational workshop.* Billings, Mont: 1989, pp. 121–129; Busby. 1979; Smith. 1979.

146. T.A. Wesche, C.M. Goertler, and C.B. Frye. "Importance and evaluation of instream and riparian cover in smaller trout streams." In *Riparian ecosystems and their management: reconciling conflicting uses.* First North American Riparian Conference, U.S. Forest Service Gen. Tech. Rep. RM-120, 1985, pp. 325–328.

147. R.J. Stuben. "Trout habitat, abundance, and fishing opportunities in fenced vs. unfenced riparian habitat along Sheep Creek, Colorado." In *Riparian ecosystems and their management: reconciling conflicting uses.* First North American Riparian Conference, U.S.D.A. Forest Service General Technical Report RM-120, 1985, pp. 310–314.

148. R.J. Behnke and M. Zarn. *Biology and management of threatened and endangered western trouts.* Ft. Collins, Colo.: U.S.F.S. General Technical Report RM-28, Rocky Mountain Forest and Range Experiment Station, 1976.

149. R. Kattelmann. "A review of watershed degradation and rehabilitation throughout the Sierra Nevada." In J.J. McDonell, J.B. Stribling, L.R. Neville, and D.J. Leopold, eds., *Watershed restoration management: physical, chemical, and biological considerations.* Herndon, Va.: American Water Resources Association, 1996, pp. 199–207.

150. S.R. Johnson, H.L. Gary, and S.L. Ponce. *Range cattle impacts on stream water quality in the Colorado Front Range.* U.S.D.A. Forest Service Research Note RM-359, 1978; D.F. Whigham, C. Chitterling, and B. Palmer. "Impacts of freshwater wetlands on water quality: a landscape perspective." *Environmental Management* 12 (1988): 663–671; D.Y. Panayotou. *Seasonal patterns of nitrate-nitrogen in the Sheep Creek riparian corridor.* Ft. Collins: Unpublished M.S. thesis, Colorado State University, 1992; J.R. Rhodes, C.M. Skau, D. Greenlee, and D.L. Brown. "Quantification of nitrate uptake by riparian forests and wetlands in an undisturbed headwaters watershed." In *Riparian ecosystems and their management: reconciling conflicting uses.* First North American Riparian Conference, U.S.D.A. Forest Service General Technical Report RM-120, 1985, pp. 175–179; T.D. Waite. *Principles of water quality.* New York. 1984; M.C. Goldberg. "Sources of nitrogen in water supplies." In T.E. Willrich and G.E. Smith, eds., *Agricultural practices and water quality.* Ames: Iowa State University Press, 1970, pp. 94–124; U.S. Environmental Protection Agency. *Water quality and pollutant source monitoring: field and laboratory procedure training manual.* Cincinnati, Ohio: EPA-430/1–76-010. U.S.E.P.A., 1976.

151. Gary, Johnson, and Ponce. 1983.

152. R.C. Szaro, S.C. Belfit, J.K. Aitkin, and J.N. Rinne. "Impact of grazing on a riparian garter snake." In *Riparian ecosystems and their management: reconciling conflicting uses.* First North American Riparian Conference, U.S.D.A. Forest Service General Technical Report RM-120, 1985, pp. 359–363.

153. S.P. Cross. "Responses of small mammals to forest riparian perturbations." In *Riparian ecosystems and their management: reconciling conflicting uses.* First North American Riparian Conference, U.S.D.A. Forest Service General Technical Report RM-120, 1985, pp. 269–275.

154. C. Keller, L. Anderson, and P. Tappel. "Fish habitat changes in Summit Creek, Idaho, after fencing the riparian area." In O.B. Cope, ed., *Forum—grazing and riparian/stream ecosystems.* Trout Unlimited, Inc., 1979, pp. 46–52; R. Van Velson. "Effects of livestock grazing upon rainbow trout in Otter Creek, Nebraska." In

O.B. Cope, ed., *Forum—grazing and riparian/stream ecosystems.* Trout Unlimited, Inc., 1979, pp. 53–55; T.J. Myers and S. Swanson. "Temporal and geomorphic variations of stream stability and morphology: Mahogany Creek, Nevada." *Journal of the American Water Resources Association* 32 (1996): 253–265; T.J. Myers and S. Swanson. "Long-term aquatic habitat restoration: Mahogany Creek, Nevada, as a case study." *Journal of the American Water Resources Association* 32 (1996): 241–252; Kauffman and Krueger. 1984.

155. W.H. Babcock. "Tenmile Creek: a study of stream relocation." *Water Resources Bulletin* 22 (1986): 405–415. [quote from p.415]

Chapter 4: The Present, and the Future

1. Colorado Demographic Information Service, Nov. 1993.

2. *Big Thompson River, Colorado Wild and Scenic River Study.* National Park Service, 1979.

3. M.J. Eubanks. *An evaluation of the Cache la Poudre Wild and Scenic River draft environmental impact statement and study report.* Ft. Collins: Colorado Water Resources Research Institute, Colorado State University, 1980.

4. *Designating segments of the Cache la Poudre River in the State of Colorado as a component of the National Wild and Scenic River System.* 99th Congress, 2d Session, Senate Report 99–354, 1986; Eubanks. 1980. (see note 3)

5. *Colorado.* Department of Highways, State of Colorado map, 1987.

6. R.R. Harris, R.J. Risser, and C.A. Fox. "A method for evaluating streamflow discharge-plant species occurrence patterns on headwater streams." In R.R. Johnson, C.D. Ziebell, D.R. Patton, P.F. Ffolliott, and R.H. Hamre, eds., *Riparian ecosystems and their management: reconciling conflicting uses.* U.S.D.A. Forest Service General Technical Report RM-120, 1985, pp. 87–90; R.L. Phipps. "Simulation of wetlands forest vegetation dynamics." *Ecological Modelling* 7 (1979): 257–288; E.H. Franz and F.A. Bazzaz. "Simulation of vegetation response to modified hydrologic regimes: a probabilistic model based on niche differentiation in a floodplain forest." *Ecology* 58 (1977): 176–183; J.C. Stromberg and D.T. Patten. "Instream flow requirements for riparian vegetation." In *Legal, institutional, financial, and environmental aspects of water issues.* American Society of Civil Engineers Proceedings, University of Delaware, 1989, pp. 123–130; J.C. Stromberg and D.T. Patten. "Riparian vegetation instream flow requirements: a case study from a diverted stream in the eastern Sierra Nevada, California, USA." *Environmental Management* 14 (1990): 185–194.

7. S.J. Shupe. "Keeping the waters flowing: stream flow protection programs, strategies and issues in the West." In L.J. MacDonnell, T.A. Rice, and S.J. Shupe, eds., *Instream flow protection in the West.* Boulder: Natural Resources Law Center, University of Colorado, 1989, pp. 1–21.

8. D.M. Gillilan and T.C. Brown. *Instream flow protection: seeking a balance in western water use.* Washington, D.C.: Island Press, Washington, 1997; S.J. Shupe. "Colorado's instream flow program: protecting free-flowing streams in a water consumptive state." In L.J. MacDonnell, T.A. Rice, and S.J. Shupe, eds., *Instream flow protection in the West.* Boulder: Natural Resources Law Center, University of Colorado, 1989, pp. 237–252; Shupe. 1989.

9. Gillilan and Brown. 1997.

10. P.R. Wilcock, G.M Kondolf, W.V.G. Matthews, and A.F. Barta. "Specification of sediment maintenance flows for a large gravel-bed river." *Water Resources Research*

32 (1996): 2911–2921; G.M. Kondolf and P.R. Wilcock. "The flushing flow problem: defining and evaluating objectives." *Water Resources Research* 32 (1996): 2589–2599.

11. J.E. O'Connor, R.H. Webb, and V.R. Baker. "Paleohydrology of pool-and-riffle pattern development: Boulder Creek, Utah." *Geological Society of America Bulletin* 97 (1986): 410–420; M.D. Harvey, R.A. Mussetter, and E.J. Wick. "A physical process-biological response model for spawning habitat formation for the endangered Colorado squawfish." *Rivers* 4 (1993): 114–131.

12. M.G. Wolman and J.P. Miller. "Magnitude and frequency of forces in geomorphic processes." *Journal of Geology* 68 (1960): 54–74; M.G. Wolman and R. Gerson. "Relative scales of time and effectiveness of climate in watershed geomorphology." *Earth Surface Processes and Landforms* 3 (1978): 189–203.

13. E.E. Wohl. "Bedrock benches and boulder bars: floods in the Burdekin Gorge of Australia." *Geological Society of America Bulletin* 104 (1992): 770–778; M.P. Collier, R.H. Webb, and E.D. Andrews. "Experimental flooding in Grand Canyon." *Scientific American* 276 (1997): 66–73.

14. N. Gordon. *Summary of technical testimony in the Colorado Water Division 1 Trial.* U.S. Forest Service General Technical Report RM-GTR-270, 1995.

15. F.R. Hauer and J.A. Stanford. "Ecology and life histories of three net-spinning caddisfly species (Hydropsychidae: Hydropsyche) in the Flathead River, Montana." *Freshwater Invertebrate Biology* 1 (1982): 18–29; J.V. Ward and J.A. Stanford. "Thermal responses in the evolutionary ecology of aquatic insects." *Annual Review of Entomology* 27 (1982): 97–117; J.A. Gore. "Hydrological change." In P. Calow and G.E. Petts, eds., *The rivers handbook: hydrological and ecological principles,* v. 2. Oxford: Blackwell Scientific Publications, 1994, pp. 33–54; H.J. Zimmermann and J.V. Ward. "A survey of regulated streams in the Rocky Mountains of Colorado, USA." In A. Lillehammer and S.J. Saltveit, eds., *Regulated rivers.* Oslo: Universitetsforlaget AS, 1984, pp. 251–262; P.D. Armitage. "Environmental changes induced by stream regulation and their effect on lotic macroinvertebrate communities." In A. Lillehammer and S.J. Saltveit, eds., *Regulated rivers.* Oslo: Universitetsforlaget AS, 1984, pp. 139–165; R.R. Harris, R.J. Risser, and C.A. Fox. "A method for evaluating streamflow discharge-plant species occurrence patterns on headwater streams." In *Riparian ecosystems and their management: reconciling conflicting uses.* First North American Riparian Conference, U.S.D.A. Forest Service General Technical Report RM- 120, 1985, pp. 87–90; J.C. Stromberg and D.T. Patten. "Riparian vegetation instream flow requirements: a case study from a diverted stream in the eastern Sierra Nevada, California, USA." *Environmental Management* 14 (1990): 185–194; L.D. Cline and J.V. Ward. "Biological and physicochemical changes downstream from construction of a subalpine reservoir, Colorado, USA." In A. Lillehammer and S.J. Saltveit, eds., *Regulated rivers.* Oslo: Universitetsforlaget AS, 1984, pp. 233–243.

16. Gillilan and Brown. 1997.

17. Shupe. 1989.

18. C.B. Stalnaker. "Evolution of instream flow habitat modeling." In P. Calow and G.E. Petts, eds., *The rivers handbook: hydrological and ecological principles,* v. 2. Oxford: Blackwell Scientific Publications, 1994, pp. 276–286; B.L. Lamb. "Quantifying instream flows: matching policy and technology." In L.J. MacDonnell, T.A. Rice, and S.J. Shupe, eds., *Instream flow protection in the West.* Boulder: Natural Resources Law Center, University of Colorado, 1989, pp. 23–39.

19. R.F. Raleigh, L.D. Zuckerman, and P.C. Nelson. *Habitat suitability index models and instream flow suitability curves: brown trout.* U.S. Fish and Wildlife Service Biological Report 82 (10.124), 1986; K.D. Bovee. *Development and evaluation of habi-*

tat suitability criteria for use in the instream flow incremental methodology. U.S. Fish and Wildlife Service Biological Report 86(7), Instream Flow Information Paper No. 21, 1986.

20. R.T. Milhous, D.L. Wegener, and T. Waddle. *User's guide to the Physical Habitat Simulation System.* Instream Flow Information Paper 11, U.S. Fish and Wildlife Service FWS/OBS-8/43 Revised, 1984; Stalnaker. 1994; Raleigh, Zuckerman, and Nelson. 1986.

21. T.J. Waddle. *A method for instream flow water management.* Ft. Collins: Unpublished Ph.D. dissertation, Colorado State University, 1992.

22. N.R.B. Olsen and S. Stokseth. "Three-dimensional modelling of hydraulic habitat in rivers with large bed roughness." In *Proceedings of the First International Symposium on Habitat Hydraulics.* The Norwegian Institute of Technology, Trondheim, Norway, Aug. 18–20, 1994, pp. 99–112; N.R.B. Olsen and K.T. Alfredsen. "A three-dimensional numerical model for calculation of hydraulic conditions for fish habitat." In *Proceedings of the First International Symposium on Habitat Hydraulics.* The Norwegian Institute of Technology, Trondheim, Norway, Aug. 18–20, 1994, pp. 113–122; A. Ghanem, P. Steffler, F. Hicks, and C. Katopodis. "Two-dimensional finite element flow modeling of physical fish habitat." In *Proceedings of the First International Symposium on Habitat Hydraulics.* The Norwegian Institute of Technology, Trondheim, Norway, Aug. 18–20, 1994, pp. 84–98; C.B. Stalnaker, K.D. Bovee, and T.J. Waddle. "The importance of the temporal aspects of habitat hydraulics in fish populations." In *Proceedings of the First International Symposium on Habitat Hydraulics.* The Norwegian Institute of Technology, Trondheim, Norway, Aug. 18–20, 1994, pp. 1–11; G.E. Petts and I. Maddock. "Flow allocation for in-river needs." In P. Calow and G.E. Petts, eds., *The rivers handbook: hydrological and ecological principles,* v. 2. Oxford: Blackwell Scientific Publications, 1994, pp. 289–307; R.J. Behnke. *Critique of instream flow methodologies.* Unpublished report for Bureau of Reclamation, 1986; Lamb. 1989.

23. J.W. Terrell, ed. *Proceedings of a workshop on fish habitat suitability index models.* U.S. Fish and Wildlife Service, WELUT Biological Report 85(6), 1984; R.J. Behnke. The illusion of technique and fisheries management, 3 pp. February 3, 1995, personal communication; R.J. Behnke. *Native trout of western North America.* Bethesda, Md.: American Fisheries Society Monograph 6, American Fisheries Society, 1992; Behnke. 1986.

24. R. Kattelmann. "A review of watershed degradation and rehabilitation throughout the Sierra Nevada." In J.J. McDonell, J.B. Stribling, L.R. Neville, and D.J. Leopold, eds., *Watershed restoration management: physical, chemical, and biological considerations.* Herndon, Va.: American Water Resources Association, 1996, pp. 199–207.

25. J.R. Vestal. "Pollution effects of storm-related runoff." In, F.E. Guthrie and J.J. Perry, eds., *Introduction to environmental toxicology.* New York: Elsevier, 1980, pp. 450–456; D.O. Doehring and M.E. Smith. *Modelling the dynamic response of floodplains to urbanization in eastern New England.* Ft. Collins: Completion Report OWRT (Office of Water Research and Technology) Project No. B-147-COLO, Environmental Resources Center, Colorado State University, 1978; D.G. Anderson. *Effects of urban development on floods in northern Virginia.* U.S. Geological Survey Water Supply Paper 2001-C, 1970; D.E. Walling. "Hydrological processes." In K.J. Gregory and D.E. Walling, eds., *Human activity and environmental processes.* Chichester: Wiley, 1987, pp. 53–85; K.S. Richards and R. Wood. "Urbanization, water redistribution, and their effects on channel processes." In K.J. Gregory, ed., *River channel changes.* Chichester: Wiley, 1977, pp. 369–388; M.D. Harvey, C.C. Watson, and S.A. Schumm. "Channelized streams: an analog for the effects of urbanization." In *Proceedings, 1983 international symposium on urban*

hydrology, hydraulics, and sediment control, 1983, pp. 401–409; L.B. Leopold. *Hydrology for urban land planning—a guidebook on the hydrologic effects of urban land use*. U.S. Geological Survey Circular No. 554, 1968; W.H. Espey, Jr. and D.E. Winslow. 1974. "Urban flood frequency characteristics." *American Society of Civil Engineers Proceedings, Hydraulics Division* 2 (1974): 279–294; M. Morisawa and E. Laflure. "Hydraulic geometry, stream equilibrium and urbanization." In D.D. Rhodes and G.P. Williams, eds., *Adjustments of the fluvial system*. Dubuque, Iowa: Kendall/Hunt Publishing Company, 1979, pp. 333–350.

26. M.G. Wolman and A.P. Schick. "Effects of construction on fluvial sediment, urban and suburban areas of Maryland." *Water Resources Research* 3 (1967): 451–464; M.G. Wolman. "A cycle of sedimentation and erosion in urban river channels." *Geografiska Annaler* 49A (1967): 385–395; Vestal. 1980.

27. Vestal. 1980.

28. S.M. Morrison. *Surveillance data, plains segment of the Cache la Poudre River, Colorado, 1970–1977*. Ft. Collins: Colorado Water Resources Research Institute, Colorado State University, 1978.

29. *A review of the U.S. Geological Survey National Water Quality Assessment pilot program*. Washington, D.C.: National Research Council, National Academy Press, 1990.

30. K.F. Dennehy, D.W. Litke, C.M. Tate, S.L. Qi, P.B. McMahon, B.W. Bruce, R.A. Kimbrough, and J.S. Heiny. *Water quality in the South Platte River basin, Colorado, Nebraska, and Wyoming, 1992–95*. U.S. Geological Survey Circular 1167, 1998.

31. G.A. Burton, Jr. "Plankton, macrophyte, fish, and amphibian toxicity testing of freshwater sediments." In G.A. Burton, Jr., ed., *Sediment toxicity assessment*. Boca Raton, Fla.: Lewis Publishers, 1992, pp. 167–182; G.A. Burton, Jr., M.K. Nelson, and C.G. Ingersoll. "Freshwater benthic toxicity tests." In G.A. Burton, Jr., ed., *Sediment toxicity assessment*. Boca Raton, Fla.: Lewis Publishers, 1992, pp. 213–240; E.A. Power and P.M. Chapman. "Assessing sediment quality." In G.A. Burton, Jr., ed., *Sediment toxicity assessment*. Boca Raton, Fla.: Lewis Publishers, 1992, pp. 1–18.

32. V.E. Forbes and T.L. Forbes. 1994. *Ecotoxicology in theory and practice*. London: Chapman and Hall, 1994; W.H. Benson and R.T. DiGiulio. "Biomarkers in hazard assessments of contaminated sediments." In G.A. Burton, Jr., ed., *Sediment toxicity assessment*. Boca Raton, Fla.: Lewis Publishers, 1992, pp. 241–265; M.L. Richardson. "Epilogue." In M. Richardson, ed., *Ecotoxicology monitoring*. Weinheim, Germany: VCH Verlagsgesellschaft, 1993, pp. 335–343; T.W. LaPoint and J.F. Fairchild. "Evaluation of sediment contaminant toxicity: the use of freshwater community structure." In G.A. Burton, Jr., ed., *Sediment toxicity assessment*. Boca Raton, Fla.: Lewis Publishers, 1992, pp. 87–110.

33. J.L. Metcalfe-Smith. "Biological water-quality assessment of rivers: use of macroinvertebrate communities." In P. Calow and G.E. Petts, eds., *The rivers handbook: hydrological and ecological principles*, v. 2. Oxford: Blackwell Scientific Publications, 1994, pp. 144–170.

34. Forbes and Forbes. 1994.

35. J.W. Hart and N.J. Jensen. "Integrated risk assessment or integrated risk management?" *Regul. Toxicol. Pharmacol.* 15 (1992): 32–40; Forbes and Forbes. 1994.

36. A.J. Underwood. "Spatial and temporal problems with monitoring." In P. Calow and G.E. Petts, eds., *The rivers handbook: hydrological and ecological principles*, v. 2. Oxford: Blackwell Scientific Publications, 1994, pp. 101–123.

37. M. Dynesius and C. Nilsson. "Fragmentation and flow regulation of river sys-

tems in the northern third of the world." *Science* 266 (1994): 753–762; M. Barinaga. "A recipe for river recovery?" *Science* 273 (1996): 1648–1650.

38. U.S. Department of Transportation. *Restoration of fish habitat in relocated streams.* Washington, D.C.: Federal Highway Administration, 1979. [quote from p. 17]

39. W.L. Jackson and B.P. Van Haveren. "Design for a stable channel in coarse alluvium for riparian zone restoration." *Water Resources Bulletin* 20 (1984): 695–703.

40. W.H. Babcock. "Tenmile Creek: A study of stream relocation." *Water Resources Bulletin* 22 (1986): 405–415.

41. D.W. Crumpacker. "The Boulder Creek corridor projects: riparian ecosystem management in an urban setting." In *Riparian ecosystems and their management: reconciling conflicting uses.* First North American Riparian Conference, 1985, pp. 389–392.

42. F.J. Swanson, T.K. Kratz, N. Caine, and R.G. Woodmansee. "Landform effects on ecosystem patterns and processes." *BioScience* 38 (1988): 92–98.

43. J.T. Pinkerton. *Knights of the broadax: the story of the Wyoming tie hack.* Caldwell, Idaho: The Caxton Printers, Ltd., 1981.

44. R.J. Behnke. "Fish culture and nonindigenous organisms." *Proceedings, symposium on non-indigenous species in western aquatic ecosystems.* Portland State University, March 27–29, 1996, pp. 15–20.

45. M. Ridley and B.S. Low. "Can selfishness save the environment?" *The Atlantic Monthly,* 1993, pp. 76–86.

46. S. Howe, personal communication, April 2000.

47. B. Messerli and J.D. Ives (eds.). *Mountains of the world: a global priority.* London: The Parthenon Publishing Group, 1997; Morrison. 1978.

Index